Microbiology

NOTICE

Medicine is an ever-changing science. As new research and clinical experience broaden our knowledge, changes in treatment and drug therapy are required. The editors and the publisher of this work have made every effort to ensure that the drug dosage schedules herein are accurate and in accord with the standards accepted at the time of publication. Readers are advised, however, to check the product information sheet included in the package of each drug they plan to administer to be certain that changes have not been made in the recommended dose or in the contraindications for administration. This recommendation is of particular importance in regard to new or infrequently used drugs.

Microbiology:
PreTest® Self-Assessment and Review
Second Edition

Edited by
Richard C. Tilton, Ph.D.

Director, Microbiology Division
Professor of Laboratory Medicine
University of Connecticut School of Medicine
Farmington, Connecticut

McGraw-Hill Book Company
Health Professions Division
PreTest Series

New York   St. Louis   San Francisco
Auckland   Bogotá   Guatemala   Hamburg
Johannesburg   Lisbon   London   Madrid
Mexico   Montreal   New Delhi   Panama
Paris   Saõ Paulo   Singapore   Sydney
Tokyo   Toronto

Library of Congress Cataloging in Publication Data
Main entry under title:

Microbiology: PreTest self-assessment and review.

First ed. (1976) edited by J. W. Foster, Jr.
Bibliography: p.
I. Medical microbiology—Examinations, questions,
etc. I. Tilton, Richard C. [DNLM: 1. Microbiology—
Examination questions. QW18 M626]
QR61.5.M57 1979   616.01'075   79-8372
ISBN 0-07-050966-2

3 4 5 6 7 8 9 0 HUHU 8 7 6 5 4 3 2 1

Editors: *Mary Ann C. Sheldon, Mark Schultz*
Editorial Assistant: *Donna Altieri*
Production Supervisor: *Susan A. Hillinski*
Production Assistant: *Judith M. Raccio*
Designer: *Robert Tutsky*
Printer: *Hull Printing Company*

Contents

Introduction

Microbiology: PreTest Self-Assessment and Review has been designed to provide medical students, as well as physicians, with a comprehensive and convenient instrument for self-assessment and review within the field of microbiology. The 500 questions provided have been designed to parallel the format and degree of difficulty of the questions contained in Part I of the National Board of Medical Examiners examinations, the Federation Licensing Examination (FLEX), the Visa Qualifying Examination, and the ECFMG examination.

Each question in the book is accompanied by an answer, a paragraph explanation, and a specific page reference to either a current journal article, a textbook, or both. A three page bibliography, listing all the sources used in the book, follows the last chapter.

Perhaps the most effective way to use this book is to allow yourself one minute to answer each question in a given chapter; as you proceed, indicate your answer beside each question. By following this suggestion, you will be approximating the time limits imposed by the board examinations previously mentioned.

When you finish answering the questions in a chapter, you should then spend as much time as you need verifying your answers and carefully reading the explanations. Although you should pay special attention to the explanations for the questions you answered incorrectly, you should read **every** explanation. The authors of this book have designed the explanations to reinforce and supplement the information tested by the questions. If, after reading the explanations for a given chapter, you feel you need still more information about the material covered, you should consult and study the references indicated.

This book meets the criteria established by the AMA's Department of Continuing Medical Education for up to 22 hours of credit in category 5D for the Physician's Recognition Award. It should provide an experience that is instructive as well as evaluative; we also hope that you enjoy it. We would be very happy to receive your comments.

Microbiology

Virology

DIRECTIONS: Each question below contains five suggested answers. Choose the **one best** response to each question.

1. Which of the following statements about cell transformation is NOT true?

(A) Cell transformation can occur spontaneously
(B) Infection with an oncogenic virus increases the rate of cell transformation
(C) Transformed cells are rounder in shape than untransformed cells
(D) Transformed cells are less subject to contact inhibition than untransformed cells
(E) Transformed cells derived from the same cell line are morphologically identical

2. Which of the following procedures CANNOT be used to synchronize randomly growing populations of cells?

(A) Bleomycin block
(B) Hydroxyurea block
(C) Double-thymidine block
(D) Hyperbaric nitrous-oxide treatment
(E) Selective detachment of mitotic cells

3. Which of the following conditions reversibly arrests growing cells in the G_1 phase of the mitotic cycle?

(A) Hydroxyurea supplementation
(B) Colcemide supplementation
(C) Double-thymidine block
(D) Isoleucine deprivation
(E) Hyperbaric nitrous-oxide treatment

4. All of the following properties are characteristic of untransformed fibroblastic cells in tissue culture EXCEPT

(A) orientation of cells in a regular, parallel arrangement
(B) balanced haploid karyotype
(C) resistance to the formation of multilayers
(D) requirement of an animal serum for growth
(E) contact inhibition of movement

5. Which of the following cell types is able to replicate in azaguanine-supplemented media?

(A) Thymidine kinase-deficient 3T3 cell
(B) Wild-type HeLa cell
(C) Lesch-Nyhan human skin fibroblast
(D) 5-Bromodeoxyuridine-resistant BRL cell
(E) Normal amniotic-fluid fibroblast

6. A new disease agent has been isolated. Which of the following procedures would be most useful in the classification of this agent as a virus or a bacterium?

(A) Protein analysis
(B) Lipid analysis
(C) Determining its nucleic acid composition
(D) Determining its filterability
(E) Determining whether it is an obligate intracellular parasite

7. A virus has been isolated from a skin lesion. In addition to producing cytopathogenicity in tissue culture, the virus is shown to be susceptible to ether and to be photodynamically inactivated following exposure to the dye neutral red. This virus is most likely to be classified as

(A) herpesvirus
(B) poxvirus
(C) adenovirus
(D) reovirus
(E) poliovirus

8. Certain plant lectins (antibody-like proteins) are mitogenic and bind to specific receptors on the cell surface of animal tissues. Included in this class of lectins is

(A) concanavalin A
(B) cytochalasin B
(C) colcemide
(D) colchicine
(E) vinblastine

9. The envelope of animal viruses can be described by all of the following statements EXCEPT

(A) it is composed of proteins
(B) it is composed of carbohydrates
(C) it is composed of lipids
(D) it is an invagination of host-cell membranes
(E) its composition can vary among virions of the same virus type

10. A tube of monkey kidney cells is inoculated with nasopharyngeal secretions. During the next seven days, no cytopathic effects (CPE) are observed. On the eighth day, however, the tissue culture is infected accidentally with a picornavirus; still, the culture does not develop CPE. The most likely explanation of this phenomenon is that

(A) picornavirus does not produce CPE
(B) the nasopharyngeal secretions contained hemagglutinins
(C) the nasopharyngeal secretions contained rubella virus
(D) picornavirus does not replicate in monkey kidney cells
(E) monkey kidney cells are resistant to CPE

11. Which of the following agents would have the LEAST effect on a virus in tissue culture?

(A) Actinomycin D
(B) 5-Fluorodeoxyuridine (FUDR)
(C) 5-Iododeoxyuridine (IUDR)
(D) Chloramphenicol
(E) Cytosine arabinoside

12. The role of the immune response in the pathology of some viral diseases is best demonstrated by which of the following viruses?

(A) Papilloma virus
(B) Hepatitis B virus
(C) Lymphocytic choriomeningitis virus
(D) Poliovirus
(E) Coxsackievirus

13. The simultaneous administration of two live-virus vaccines can result in

(A) enhancement of the immune response to both vaccines
(B) inhibition of the immune response to one of the vaccines
(C) an increased risk of immunologic complications
(D) an antibody response mediated only by immunoglobulin G
(E) mutation of one of the viruses to the virulent form

14. Which of the following statements best describes the suspected mode of action of interferon in producing resistance to viral infection?

(A) It stimulates cell-mediated immunity
(B) It stimulates humoral immunity
(C) Its direct antiviral action is related to the suppression of messenger RNA formation
(D) Its action is related to the synthesis of a protein inhibitor of translation or transcription
(E) It alters the permeability of the cell membrane so that viruses cannot enter the cell

15. A vaccine is available for rubella. Administration of this vaccine would be ill-advised for all of the following individuals EXCEPT

(A) a pregnant woman
(B) a woman who intends to become pregnant next month
(C) a prepubertal girl
(D) an individual who is allergic to eggs
(E) an individual who has an immune deficiency disease

16. A medium-sized DNA virus with a lipid-containing envelope around the viral capsid would most likely belong to which of the following groups?

(A) Poxvirus
(B) Herpesvirus
(C) Adenovirus
(D) Enterovirus
(E) Reovirus

17. Picornaviruses are small (20 to 30 nm), single-stranded, RNA-containing viruses. Which of the following viruses is NOT a picornavirus?

(A) Poliovirus
(B) Coxsackievirus A
(C) Echovirus type 8
(D) Rhinovirus
(E) Parainfluenza virus

18. The mode of transmission of viruses is an important feature in the understanding of viral disease processes. Which of the following diseases is caused by a virus spread primarily by the fecal-oral route?

(A) Poliomyelitis
(B) Dengue
(C) Yellow fever
(D) St. Louis encephalitis
(E) Japanese B encephalitis

19. Which of the following pairs of viruses contains two taxonomically **unrelated** organisms?

(A) Rabies virus and Marburg virus
(B) Herpes simplex virus and simian B virus
(C) Mumps virus and parainfluenza virus type 4
(D) Influenza virus and Rous sarcoma virus
(E) Polyoma virus and simian virus 40 (SV40)

20. Herpesviruses have been associated with all of the following disease processes EXCEPT

(A) chickenpox
(B) cytomegalic inclusion disease
(C) Kaposi's varicelliform eruption
(D) shingles
(E) verrucae

21. Mumps virus is biologically related to the virus causing which of the following diseases?

(A) Rabies
(B) Hepatitis A
(C) Measles
(D) Variola
(E) Varicella

22. Reoviruses are double-stranded-RNA viruses that

(A) cause severe respiratory disease
(B) cause severe gastrointestinal disease
(C) are respiratory and enteric "orphans"
(D) are recovered rarely from healthy individuals
(E) are unique to man and certain other primates

23. Echoviruses are cytopathogenic human viruses that mainly infect the

(A) respiratory system
(B) central nervous system
(C) blood and lymphatic systems
(D) intestinal tract
(E) bladder and urinary tract

24. Which of the following statements is true of all tumor (oncogenic) viruses?

(A) They contain double-stranded DNA
(B) They produce sarcomas
(C) They change the growth properties of infected cells
(D) They exist as episomes in the cytoplasm of infected cells
(E) They are transmitted by insect vectors

25. Epidemic pleurodynia and myocarditis of newborn infants both are caused by

(A) group B coxsackievirus
(B) polyoma virus
(C) respiratory syncytial (RS) virus
(D) reovirus
(E) cytomegalovirus

26. All of the following statements pertain to exanthem subitum (roseola infantum) EXCEPT

(A) it is occasionally fatal
(B) it has an incubation period of 10 to 14 days
(C) it is associated with lymphadenopathy
(D) it is associated with a rubelliform rash
(E) it usually occurs in infants between six months and three years of age

27. Warts are caused most frequently by which of the following viruses?

(A) Adenovirus
(B) Picornavirus
(C) Polyoma virus
(D) Papilloma virus
(E) Herpesvirus

28. Treatment of influenza virus with 1M magnesium sulfate followed by heating at 50°C for 30 minutes would be expected to

(A) kill the virus
(B) destroy infectivity but maintain antigenicity
(C) destroy antigenicity and infectivity
(D) have no effect on the infectivity of the virus
(E) stimulate viral replication

29. After returning from Africa, a woman develops an illness characterized by high fever, black vomitus, proteinuria, and jaundice. She is most likely to have

(A) dengue
(B) yellow fever
(C) kala-azar
(D) scrub typhus
(E) equine encephalitis

30. Antigenic skin testing can be used to help establish a diagnosis for all of the following infectious disorders EXCEPT

(A) lymphogranuloma venereum
(B) sarcoidosis
(C) herpes
(D) cat scratch fever
(E) poliomyelitis

31. Cytomegaloviruses are typical herpesviruses and can cause disease that varies in severity from very mild to life-threatening, depending on the host. Which of the following individuals would pose the greatest hazard for transmission of cytomegalovirus (CMV) to hospital personnel?

(A) A 50-year-old patient on the surgical service whose CMV titer is 1:80
(B) A three-week-old baby who has congenital abnormalities
(C) An adolescent female recently vaccinated for rubella
(D) A new mother who complained of malaise in the second trimester of pregnancy
(E) The hospital virologist

32. Which of the following viruses is NOT associated with latent infections?

(A) Lymphocytic choriomeningitis virus
(B) Adenovirus
(C) Smallpox virus
(D) Herpes simplex virus
(E) Herpes zoster virus

33. Which of the following viruses usually causes disseminated infection?

(A) Human papilloma virus
(B) Poliovirus
(C) Rhinovirus
(D) Influenza virus
(E) Parainfluenza virus

34. Meningitis is characterized by the acute onset of fever and stiff neck. "Aseptic meningitis" may be caused by a variety of microbial agents. During the initial 24 hours of the course of aseptic meningitis, an affected individual's cerebrospinal fluid is characterized by

(A) decreased protein content
(B) elevated glucose concentration
(C) lymphocytosis
(D) polymorphonuclear leukocytosis
(E) eosinophilia

35. A woman develops "hepatitis" two months after she was transfused with 10 units of whole blood. She is found to have neither HB_sAg nor anti-HB_sAg. The most likely cause of her disease is

(A) hepatitis A virus
(B) non-A, non-B hepatitis virus
(C) cytomegalovirus
(D) influenza virus
(E) delayed transfusion reaction to incompatible antigens

36. Rabies is a disease that can have dire consequences in an infected human. Which of the following animals poses the most significant rabies threat to residents of the United States?

(A) Bats
(B) Dogs
(C) Cats
(D) Rats
(E) Mice

37. Which of the following statements does NOT pertain to simian herpesviruses?

(A) They also are known as SV40 viruses
(B) They have oncogenic capacity in monkeys
(C) They are nononcogenic for humans
(D) They are extremely virulent in humans
(E) They have a DNA genome

38. Individuals who have had varicella as children occasionally suffer a recurrent form of the disease, called shingles, as adults. The agent causing these diseases is a member of which of the following viral families?

(A) Herpesvirus
(B) Poxvirus
(C) Adenovirus
(D) Myxovirus
(E) Paramyxovirus

39. The common cold is caused most often by

(A) adenovirus
(B) influenza virus
(C) respiratory syncytial virus
(D) rhinovirus
(E) parainfluenza virus

40. A virus that produces mild disease accompanied by rash, that causes congenital anomalies in infants whose mothers were infected in the first trimester of pregnancy, and against which vaccination is effective is most likely

(A) measles virus
(B) rubella virus
(C) mumps virus
(D) varicella virus
(E) respiratory syncytial virus

41. Shortly after returning from Africa, a Peace Corps worker develops fever and pharyngitis of acute onset. A virus isolated from the individual's blood is an RNA virus that cross-reacts immunologically with lymphocytic choriomeningitis (LCM) virus. The disease described most likely is

(A) trachoma
(B) yaws
(C) Rift Valley fever
(D) Q fever
(E) Lassa fever

42. The most effective measure for the prevention of rubeola is the administration of

(A) antitoxin
(B) convalescent serum
(C) gamma globulin
(D) measles virus vaccine
(E) live attenuated measles virus vaccine

43. Hepatitis-associated surface antigen is found in

(A) the blood of many patients who have hepatitis A
(B) the feces of many patients who have hepatitis A
(C) the blood of many patients who have hepatitis B
(D) the feces of many patients who have hepatitis B
(E) the blood of many patients who have chemical hepatitis

44. Although mumps virus characteristically invades the parotid glands, this virus also commonly produces inflammation in the

(A) testes
(B) kidneys
(C) bones
(D) skin
(E) colon

45. An initial smallpox vaccination normally produces maximal local reaction (so-called "primary take") how long after inoculation?

(A) 1 to 2 days
(B) 4 to 6 days
(C) 7 to 9 days
(D) 10 to 14 days
(E) 2 to 3 weeks

46. Which of the following disorders is NOT associated with the congenital rubella syndrome?

(A) Cataracts
(B) Deafness
(C) Cleft palate
(D) Hepatosplenomegaly
(E) Patent ductus arteriosus

47. The presence of Negri inclusion bodies in host cells is characteristic of

(A) mumps
(B) infectious mononucleosis
(C) congenital rubella
(D) aseptic meningitis
(E) rabies

48. Poliomyelitis is an acute infectious disease capable of causing serious pathology of the central nervous system. Which of the following individuals, all of whom are **not** immune to polio, is at greatest risk for contracting polio?

(A) A person who, while camping, was bitten by a tick
(B) A child in a boarding school who slept in the same dormitory as a child who subsequently contracted polio
(C) An aide who washed floors on a polio ward
(D) A person who had repeated sexual intercourse with a polio victim
(E) A person who ate salad prepared by an individual diagnosed the following day as having polio

49. According to recommendations issued by the U.S. Public Health Service, which of the following statements regarding vaccination against smallpox is true?

(A) Pregnant women should be vaccinated in the first trimester
(B) Individuals who have eczema should be vaccinated soon after diagnosis
(C) Individuals who have immune deficiencies should be vaccinated every five years
(D) Individuals traveling abroad need not be vaccinated
(E) Children need not be vaccinated until they begin school

50. A woman appears in a hospital's emergency room with vesicular lesions on her vulva; these lesions clinically resemble those caused by herpes simplex virus. From which of the following clinical specimens would the virus most likely be isolated?

(A) Urine
(B) Blood
(C) Stool
(D) Throat secretions
(E) Vesicular fluid

51. A 17-year-old female college student has splenomegaly. Serologic examination reveals an elevated white blood cell count (including atypical lymphocytes) and heterophil antibodies. She probably has

(A) mumps
(B) parainfluenza
(C) rubella
(D) infectious mononucleosis
(E) lymphocytic choriomeningitis

52. Influenza virus infections can be characterized by which of the following statements?

(A) They have a high case-fatality rate
(B) They confer type-specific immunity
(C) They frequently are transmitted from infected household pets
(D) They commonly persist in human carriers
(E) They are more severe when the virus is type B, not types A or C

53. A hospital worker is found to have hepatitis-associated surface antigen. Subsequent tests reveal the presence of *e* antigen as well. This individual most likely

(A) has active hepatitis and is infective
(B) is infective but does not have active hepatitis
(C) is not infective
(D) is evincing a biologic false positive test for hepatitis
(E) has recently consumed raw shellfish

54. Which of the following clinical syndromes is most commonly associated with rhinovirus infection in an adult?

(A) Bronchitis
(B) Bronchiolitis
(C) Bronchopneumonia
(D) Croup
(E) Common cold

55. Examination of a Giemsa-stained smear from scrapings at the base of a skin lesion may help to differentiate chickenpox (varicella) from smallpox (variola). Which of the following findings characterizes smears from chickenpox lesions?

(A) Eosinophilic intranuclear inclusion bodies
(B) Eosinophilic cytoplasmic inclusion bodies
(C) Basophilic cytoplasmic inclusion bodies
(D) Large syncytial masses
(E) An absence of multinuclear giant cells

DIRECTIONS: Each question below contains four suggested answers of which **one** or **more** is correct. Choose the answer:

A	if	1, 2, and 3	are correct
B	if	1 and 3	are correct
C	if	2 and 4	are correct
D	if	4	is correct
E	if	1, 2, 3, and 4	are correct

56. Robert Koch, a nineteenth-century German bacteriologist, established criteria for recognition of pathogenic microbes. These "postulates" include the need to

(1) observe a given pathogenic microorganism in constant association with a given disease
(2) isolate and grow the microorganism in pure culture
(3) reproduce the disease in a susceptible animal by inoculation with the microorganism in a pure culture
(4) isolate the microorganism from the infected animal and show it to be identical to the microbe inoculated into the animal

57. Viruses have been alleged to be structurally simple microorganisms that are metabolically inert in the absence of a host cell. Which of the following statements about viruses are true?

(1) They contain a single strand of either RNA or DNA
(2) They depend on host cells for all enzymes
(3) They contain only protein and nucleic acid in the virion
(4) They can package cellular components

58. Interferon is a protein that inhibits virus replication. It is produced by cells in tissue culture when the cells are stimulated with

(1) viruses
(2) synthetic double-stranded polynucleotides
(3) endotoxin
(4) rickettsiae

59. In which of the following viruses does the virion consist only of the nucleocapsid and its nucleic acid core?

(1) Adenovirus
(2) Poxvirus
(3) Picornavirus
(4) Herpesvirus

60. The pathologic effects of viruses on host cells can be useful in the identification of viruses. These cytopathic effects are

(1) usually morphologic in nature
(2) usually fatal to the host cell
(3) often associated with changes in lysosomal membranes
(4) necessarily pathognomonic for an infecting virus

SUMMARY OF DIRECTIONS

A	B	C	D	E
1, 2, 3 only	1, 3 only	2, 4 only	4 only	All are correct

61. Which of the following viruses would lose their infectivity by being heated at 60°C for one hour?

(1) Measles virus
(2) Rabies virus
(3) Mumps virus
(4) Hepatitis virus A

62. Specific immunity to viral agents is directed against the

(1) surface components of the virion
(2) viral proteins synthesized within the infected cell
(3) virus-specific surface proteins in the infected cell's membrane
(4) virion-associated enzymes

63. Virus inclusion bodies are described by which of the following statements?

(1) They are composed of assembled progeny virions and host-cell constituents
(2) They are found in both the cytoplasm and the nucleus
(3) They have little effect on the metabolism of the host cell
(4) They are useful in the diagnosis of viral disease

64. The virulence of virus strains is

(1) under polygenic control
(2) related to host-cell variables
(3) correlated with their resistance to interferon
(4) correlated with their ability to multiply at host-cell temperatures

65. A virion-antibody complex may be

(1) reversibly bound
(2) irreversibly bound
(3) infectious in one type of host cell but not in another
(4) dissociated by physicochemical means

66. The use of interferon inducers may prove valuable in the control of viral disease, because interferon

(1) is effective in already infected as well as in uninfected cells
(2) has a wide spectrum of antiviral activity
(3) acts irreversibly
(4) is apparently weakly antigenic

67. Individuals who have which of the following diseases can be treated by a regimen including appropriately administered gamma globulin?

(1) Hepatitis A
(2) Hepatitis B
(3) Rabies
(4) Poliomyelitis

68. Herpesviruses have been associated with

(1) infectious mononucleosis
(2) nasopharyngeal carcinoma
(3) Burkitt's lymphoma
(4) carcinoma of the cervix

69. Subacute sclerosing panencephalitis (SSPE) is a slow-virus infection of humans. As a result of having this disease, a patient with SSPE might have antibodies capable of reacting with the virus causing

(1) kuru
(2) scrapie
(3) Creutzfeldt-Jakob disease
(4) measles

70. Viruses can be transmitted by

(1) arthropod vectors
(2) animal bites
(3) food or drink
(4) person-to-person contact

71. Which of the following statements about western equine encephalitis (WEE) are true?

(1) It is caused by a group A togavirus
(2) It is maintained in wild birds
(3) It is transmitted by mosquitoes
(4) Its primary reservoir is in humans

72. At present, routine laboratory tests are available for the detection of

(1) hepatitis-associated surface antigen (HB$_s$AG)
(2) *e* antigen
(3) anti-HB$_s$Ag
(4) HAV (hepatitis A virus)

73. Which of the following microorganisms are species-specific for the human host?

(1) Poliovirus
(2) Variola virus
(3) Lymphocytic choriomeningitis (LCM) virus
(4) Cytomegalovirus

74. Which of the following viruses can produce oncogenic disease in mice and hamsters?

(1) Arenavirus
(2) Parvovirus
(3) Picornavirus
(4) Adenovirus

75. Which of the following diseases are caused by DNA viruses?

(1) Smallpox
(2) Measles
(3) Chickenpox
(4) Yellow fever

76. Common characteristics of arboviruses include

(1) transmission by arthropod vectors
(2) resistance to ether
(3) production of encephalitic disease in humans
(4) a genome of double-stranded DNA

SUMMARY OF DIRECTIONS

A	B	C	D	E
1, 2, 3 only	1, 3 only	2, 4 only	4 only	All are correct

77. Which of the following statements about individuals suffering from infectious encephalitides are true?

(1) Their disease is more likely to be bacterial, rather than viral, in origin
(2) They usually remain afebrile
(3) They can be treated easily with intrathecal penicillin
(4) They frequently develop neurologic sequellae

78. Included in the variety of clinical syndromes caused by echovirus are

(1) respiratory and enteric disease
(2) paralysis
(3) febrile illness with rash
(4) conjunctivitis

79. Epidemic pleurodynia (Bornholm's disease) is associated with

(1) group B coxsackievirus
(2) a peak incidence in late summer
(3) abdominal pain in approximately 50 percent of cases
(4) orchitis in most affected male individuals

80. Coronaviruses are recognized by club-shaped surface projections that are 20 nm long and resemble solar coronas. These viruses are characterized by their ability to

(1) grow well in the usual laboratory cultured cell lines
(2) cause the common cold
(3) infect infants more frequently than adults
(4) grow profusely at 37°C

81. The viruses causing mumps, parainfluenza viral disease, and rubella are all paramyxoviruses. Characteristics of paramyxoviruses include

(1) the presence of a hemagglutinin or a neuraminidase
(2) hemolytic activity
(3) the ability to cause a persistent noncytocidal infection
(4) an RNA genome

82. Rabies is an acute infection of the central nervous system. Which of the following statements about the rabies virus are true?

(1) It can be isolated from the blood of infected patients
(2) It is of a single antigenic type
(3) It can be transmitted by a dog four weeks before the dog becomes noticeably ill
(4) It produces an infection that is nearly always fatal in humans

83. Infection with cytomegalo-virus can

(1) be transmitted transplacentally
(2) be activated by immunosuppres-sive agents
(3) cause stillbirths
(4) be treated effectively

84. The ability to adsorb to receptors on red blood cells is a feature of some viruses, including the paramyxoviruses. A test to demonstrate paramyxoviral hemagglutinins should be performed at 4°C, because at this temperature

(1) the cell receptor site is unmasked
(2) the activity of the hemagglutinin is preserved
(3) a specific inhibitor of hemagglu-tination is induced
(4) neuraminidase is inactivated

85. Varicella-zoster virus can be characterized by

(1) the ability to cause severe disease in immunosuppressed children
(2) formation of cross-reacting anti-bodies with herpes simplex virus type 2
(3) infection of the posterior nerve roots and ganglia of adults
(4) production of painless genital lesions

86. It has been observed that certain New Guinea tribes fond of eating un-cooked human brains suffer from a neurologic disease characterized by mental disease, ataxia, and death months after ingestion of the brains. Efforts to identify this disease have led to the

(1) isolation of a virus from the in-gested human brains
(2) detection in patients of antibodies to kuru virus
(3) immunofluorescent staining of virus particles in brain tissue
(4) experimental transmission of the disease to chimpanzees

87. Infectious mononucleosis is asso-ciated with

(1) heterophil antibodies
(2) bilateral hilar adenopathy on chest x-ray
(3) atypical lymphocytes
(4) Epstein-Barr virus

88. Disease caused by hepatitis A virus can be characterized by which of the following statements?

(1) It can be transmitted by aerosols
(2) Parenteral transmission does not occur
(3) An incubation period of 60 days is average
(4) It clinically can resemble hepa-titis B virus disease

89. Parainfluenza virus type 1 is known to

(1) lyse erythrocytes
(2) agglutinate erythrocytes
(3) cause croup in children
(4) cause bronchiectasis in the elderly

90. Coxsackieviruses comprise a large subgroup of the enteroviruses. They produce a wide variety of disorders in humans, including

(1) aseptic meningitis
(2) herpangina
(3) myocarditis (in adults)
(4) vomiting (in infants)

91. Clinical and histologic findings characteristic of smallpox include

(1) lesions in varying stages of development in a given affected area
(2) Guarnieri bodies
(3) nuclear inclusions
(4) hyperplasia of the reticuloendothelial system

92. Which of the following statements about measles (rubeola) virus infection are true?

(1) It may cause subacute sclerosing panencephalitis (SSPE)
(2) It can be diagnosed by the presence of Koplik's spots
(3) It may cause giant cell pneumonia
(4) It is complicated by encephalomyelitis in approximately 1 percent of cases

93. Molluscum contagiosum virus, a poxvirus, can be described by which of the following statements?

(1) It has been transmitted experimentally to humans
(2) It differs in appearance on electron microscopy from other poxviruses
(3) It produces proliferative lesions on genital epithelium
(4) It usually infects older persons

94. In its early stages, dengue is characterized by

(1) arthralgia
(2) wheal and flare reactions
(3) punctiform rash
(4) edema

95. Acute herpetic gingivostomatitis (Vincent's stomatitis), which is the most common primary infection with herpesvirus hominis type 1, is a disease that

(1) has an incubation period of two weeks
(2) usually does not cause fever
(3) is most common in adolescents
(4) causes regional lymphadenitis

96. A recently identified virus has the following biologic characteristics: it is a double-stranded-RNA virus; it has a double-walled capsid; and it resembles a rotavirus. Which of the following statements about this virus are valid?

(1) This virus likely is similar to Nebraska calf diarrhea virus
(2) Most newborn infants have maternally acquired rotavirus antibodies
(3) Early breast-feeding helps protect neonates against rotavirus
(4) Rotavirus is a major cause of neonatal diarrhea

97. Mumps virus accounts for 10 to 15 percent of all cases of aseptic meningitis in the United States. Infection with mumps virus is

(1) apt to recur periodically in many affected individuals
(2) a leading cause of male sterility due to mumps orchitis
(3) maintained in a large canine reservoir
(4) preventable by immunization

98. The serum of a newborn infant reveals a 1:32 cytomegalovirus (CMV) titer. The child is clinically asymptomatic. Which of the following courses of action would be advisable?

(1) Repeat the CMV titer immediately
(2) Obtain a CMV titer from all siblings
(3) Obtain an anti-CMV immunoglobulin M titer from the mother
(4) Obtain an anti-CMV IgM titer from the baby

99. Antigenic variation in influenza virus is a phenomenon that

(1) occurs by the addition of new antigenic determinants
(2) is more prominent in B strains than in A strains
(3) affects both the hemagglutinin and neuraminidase antigens
(4) has no impact on the clinical management of influenza

100. Encephalitis may be caused by a number of neurotropic viruses, among them the St. Louis encephalitis virus. Infection with this virus

(1) produces lesions that are most marked in the midbrain and brain stem
(2) has a higher incidence in children and infants than in adults
(3) has an incubation period of one to three weeks
(4) is transmitted by ticks from an avian reservoir

101. Rabies virus, which produces one of the most feared of all human diseases, is

(1) an ether-sensitive RNA virus
(2) associated with acute ascending paralysis in some cases
(3) the cause of Negri bodies in the cytoplasm of nerve cells
(4) associated with an incubation period in humans of 7 to 10 days

SUMMARY OF DIRECTIONS

A	B	C	D	E
1, 2, 3 only	1, 3 only	2, 4 only	4 only	All are correct

102. Which of the following statements about herpes simplex virus infection, one of the more common human infections, are true?

(1) It may recur repeatedly in some individuals
(2) It rarely recurs in a host who has a high antibody titer
(3) It can be reactivated by emotional disturbances or prolonged exposure to sunlight
(4) Initial infection usually occurs due to intestinal absorption of the virus

103. Certain viruses have been associated with birth defects. These teratogenic viruses include

(1) rubella virus
(2) cytomegalovirus
(3) coxsackievirus
(4) herpes simplex virus

104. Infectious mononucleosis can be a debilitating disease. This viral disorder can be characterized by which of the following statements?

(1) It is caused by a type of rhabdovirus
(2) It is most prevalent in the 10-to-15-year-old age group
(3) Affected individuals respond to treatment with heterophil antibodies
(4) The causative virus has been linked to Burkitt's lymphoma

105. Based on known mechanisms of activity, which of the following compounds might be used to treat individuals who have smallpox?

(1) Transfer factor
(2) Vaccinia hyperimmune serum
(3) N-methylisatin-β-thiosemicarbazone (methisazone)
(4) Tetracycline

DIRECTIONS: The groups of questions below consist of lettered choices followed by several numbered items. For each numbered item select the **one** lettered choice with which it is **most** closely associated. Each lettered choice may be used once, more than once, or not at all.

Questions 106-110

For each of the following viral diseases, choose the source from which the vaccine for that disease is obtained.

(A) Calf or sheep lymph
(B) Duck embryo
(C) Chick embryo cell culture
(D) Chick embryo tissue culture
(E) Monkey kidney tissue culture

106. Eastern equine encephalitis

107. Mumps

108. Measles

109. Rabies

110. Smallpox

Questions 111-114

All of the substances below are associated with hepatitis. For each substance, choose the description with which it is most likely to be associated.

(A) Detectable by radioimmunoassay (RIA)
(B) Transmitted primarily by the fecal-oral route
(C) Transmitted primarily by the parenteral route
(D) Responds well to nucleic acid analogs (ara-A, ara-C)
(E) Closely associated with DNA polymerase activity

111. $HB_s Ag$

112. HAV

113. $HB_c Ag$

114. HBV

Virology

Answers

1. The answer is E. *(Watson, ed 3. pp 558-560.)* Cell transformation is the process by which a cell with normal growth characteristics is changed into a cell that is less subject to contact inhibition of growth, less nutritionally fastidious, rounder in shape, and more capable of neoplastic development. Infection with oncogenic viruses and treatment with carcinogens or radiation can transform cells at a much higher frequency than that of spontaneous transformation, which occurs rarely. Transformed cells from the same cell line may display varying biochemical and morphologic characteristics.

2. The answer is A. *(Goodman, ed 5. pp 1290-1292. Tobey, J Cell Physiol 79 [1972]:259-265.)* Except for bleomycin block, all the methods listed in the question can interrupt the cell cycle in a reversible fashion. Bleomycin, in contrast, blocks the cell cycle irreversibly in G_2, the stage between DNA synthesis and mitosis. This antitumor agent is thought to exert its cytotoxic action by fragmentation of DNA; it can also bind to DNA and cause "nicking." Bleomycin is used primarily for the treatment of individuals who have squamous cell carcinoma, testicular carcinoma, or certain lymphomas.

3. The answer is D. *(Tobey, Cancer Res 31 [1971]:46. Watson, ed 3. pp 558-560.)* Isoleucine deprivation is the only procedure listed in the question able to block cells reversibly in G_1, the cell-cycle stage preceding DNA synthesis. Isoleucine is an essential growth factor for cell division. If cultured mammalian cells are deprived of isoleucine, arrest of growth in the G_1 stage occurs within 24 to 36 hours; subsequent addition of isoleucine prompts the cell population to initiate DNA synthesis and undergo mitosis in synchrony.

4. The answer is B. *(Davis, ed 2. pp 1122-1123, 1142.)* Tissue culture of fibroblast-like cells requires a culture medium containing animal serum and selected nutrients. As the cells grow, they spread out along the bottom of the culture flask in a regular, parallel orientation. Fibroblastic cells in culture form a monolayer on the flask bottom; contact inhibition prevents clumping and piling up of cells. Untransformed fibroblastic cells in primary culture have a diploid genotype; however, 3T3 mouse fibroblasts, used in virologic study, are aneuploid.

5. The answer is C. *(Fujimoto, Proc Natl Acad Sci 68 [1971]:1516-1519. Lehninger, ed 2. p 742.)* The Lesch-Nyhan syndrome is caused by a genetic defect in the production of the cellular enzyme hypoxanthine-guanine phosphoribosyltransferase. Normal cells, containing this enzyme, are able to incorporate the guanine analog azaguanine into nucleosides and thus produce nucleic-acid products lethal to the cells. Because Lesch-Nyhan cells lack the necessary enzyme, they can survive in an azaguanine-supplemented medium. The other cell types cited in the question all contain metabolically active hypoxanthine-guanine phosphoribosyltransferase.

6. The answer is C. *(Jawetz, ed 13. pp 311-312.)* The presence of **either** RNA **or** DNA, not both, is evidence that an unknown disease agent is of viral origin. Neither filterability nor type of host-agent relationship always distinguishes between bacteria and viruses. Certain bacteria may be smaller than large viruses and, thus, be filterable, and some bacteria, like all viruses, are obligate intracellular parasites. Bacteria and viruses both contain protein and lipid.

7. The answer is A. *(Jawetz, ed 13. p 314.)* Most viruses are inactivated by prolonged heating, extremes in pH, and radiation, including x-ray and ultraviolet radiation. Of the viruses listed in the question, only herpesvirus also is both susceptible to ether and photodynamically inactivated after exposure to vital dyes, e.g., neutral red and toluidine blue. Both ether and neutral-red treatments have been used clinically in attempts to eradicate lesions stemming from herpes simplex disease.

8. The answer is A. *(Nicolson, J Cell Biol 60 [1974]:236-248.)* Lectins are proteins whose biochemical actions resemble those of antibodies. Concanavalin A is a plant lectin that specifically binds to α-**D**-mannopyranosyl and β-**D**-galactopyranosyl residues on the surface of animal cells. It exerts a mitogenic effect on these cells. Unlike concanavalin A, colchicine and vinblastine bind to elements of the microtubular system, and colcemide and cytochalasin B attach to structures of the microfilament system.

9. The answer is D. *(Davis, ed 2. p 1031.)* The envelope of certain animal viruses is composed of proteins, glycoproteins, and lipids. Although the envelope contains substances associated with the host-cell membrane — in addition to virus-specific components — it is not an invagination of the host-cell membrane. Because virions incorporate host-cell compounds into their envelopes, the composition of the envelopes can vary from virion to virion, even within the same viral group.

10. The answer is C. *(Volk, p 551.)* Rubella virus does not produce cytopathic effects (CPE) in tissue-culture cells. Moreover, rubella-infected cells challenged with a picornavirus are resistant to subsequent infection and thus would not exhibit CPE. Mouse kidney cells infected only with picornavirus would show CPE.

11. The answer is D. *(Braude, vol 8. p 30. Davis, ed 2. pp 1183-1186.)* Actinomycin D, 5-fluorodeoxyuridine (FUDR), 5-iododeoxyuridine (IUDR), and cytosine arabinoside are among the agents that can affect the growth of viruses in culture. The common broad-spectrum antibiotics, including chloramphenicol, have no effect on viral growth. Chloramphenicol interferes with bacterial protein synthesis by inhibiting the enzyme peptidyl transferase.

12. The answer is C. *(Davis, ed 2. pp 1215-1216.)* In several viral diseases, including lymphocytic choriomeningitis, the immune response in the host manifests as a delayed hypersensitivity to the virus. This reaction is thought to be caused by an allergic response to the virus or its products. For individuals who have lymphocytic choriomeningitis, x-ray irradiation or administration of anti-lymphocytic serum can suppress the host immune response and thus reduce the clinical signs and symptoms of the disease.

13. The answer is B. *(Bellanti, pp 397-398.)* In certain combinations of live viruses, the presence of one virus may inhibit the growth of the other. If this reaction occurs when two live-virus vaccines are administered simultaneously, immunization against the inhibited virus will not be successful. This mechanism of interference may be mediated by interferon.

14. The answer is D. *(Bellanti, pp 397-398.)* Interferon is a protein produced by cells in response to a viral infection or certain other agents. Entering uninfected cells, interferon causes production of a second protein that alters protein synthesis. As a result, due to inhibition of either translation or transcription, new viruses are not assembled following infection of interferon-protected cells.

15. The answer is C. *(Jawetz, ed 13. p 427.)* Rubella virus vaccine (a live attenuated-virus vaccine) can cross the placenta of a pregnant woman and infect the fetus. For this reason, pregnancy and intended pregnancy are contraindications to the administration of this vaccine. The vaccine also may produce serious side effects in individuals allergic to chicken protein as well as in persons with abnormal immune function.

16. The answer is B. *(Davis, ed 2. p 1141.)* Herpesviruses contain a double-stranded DNA genome of 50 to 100 x 10^6 daltons, surrounded by a protein coat that is in turn enclosed by a lipid envelope. The DNA genomes of poxviruses are very large, i.e., 160 x 10^6 daltons; these viruses have lipid in the outer coat but no definite envelope. Adenoviruses, which contain DNA, are naked icosahedral structures without lipid envelopes. Reoviruses and enteroviruses are RNA-containing viruses.

17. The answer is E. *(Davis, ed 2. p 1281.)* Poliovirus, coxsackievirus, echovirus, and rhinovirus all are classified as picornaviruses. These viruses are small (20 to 30 nm), are resistant to ether, and have a genome comprised of single-stranded RNA; in addition, the nucleocapsid is naked. Parainfluenza virus is a paramyxovirus.

18. The answer is A. *(Davis, ed 2. p 1281.)* Poliovirus, like other enteroviruses in the picornavirus group, is spread mainly by fecal-oral transmission. The other diseases listed in the question are caused by arthropod-borne viruses. The viral agents of this group of diseases all are classified as togaviruses, group B.

19. The answer is D. *(Davis, ed 2. pp 1141, 1240, 1336, 1368.)* Rabies virus and Marburg virus are classified as rhabdoviruses; herpes simplex and simian B viruses, herpesviruses; mumps virus and parainfluenza virus type 4, paramyxoviruses; and polyoma virus and simian virus 40 (SV40), papovaviruses. Influenza virus is a myxovirus causing influenza and similar illnesses in humans. Rous sarcoma virus (RSV) is a leukovirus capable of causing cell transformation and oncogenic disease.

20. The answer is E. *(Jawetz, ed 13. pp 447-448.)* Herpesviruses (of which at least 25 varieties are known) contain double-stranded DNA. Among the wide variety of diseases linked to herpesvirus infection are chickenpox, shingles, cytomegalic inclusion disease, Kaposi's varicelliform eruption, labial and cervical carcinoma, lymphoma, keratoconjunctivitis, and meningoencephalitis. Verrucae, or warts, are caused by papovaviruses.

21. The answer is C. *(Davis, ed 2. p 1331.)* Both mumps and measles are diseases caused by paramyxoviruses, which are enveloped and contain single-stranded RNA. Rabies is caused by a rhabdovirus, variola (smallpox) by a poxvirus, and varicella (chickenpox) by a herpesvirus. The causative agent of hepatitis is as yet unclassified.

22. The answer is C. *(Davis, ed 2. pp 1400-1407.)* The Reoviridae family of viruses consists of the reoviruses and the orbiviruses. Reovirus (respiratory enteric "orphan" virus) frequently produces asymptomatic infection of the respiratory and gastrointestinal tracts; if clinical disease results, it is mild and self-limiting. Reoviruses infect all mammals except whales.

23. The answer is D. *(Jawetz, ed 13. pp 381-382.)* Echoviruses were discovered accidentally during studies on poliomyelitis. They were named enteric cytopathogenic human orphan (ECHO) viruses because, at the time, they had not been linked to human disease and thus were considered "orphans." Echoviruses now are known to infect the intestinal tract of humans; they also can cause aseptic meningitis, febrile illnesses, and the common cold. Echoviruses range in size from 24 to 30 nm in diameter and contain a core of RNA.

24. The answer is C. *(Davis, ed 2. pp 1419-1433.)* By definition, a tumor (oncogenic) virus changes the growth properties of cells. Tumor viruses are found in both the RNA and DNA viral groups and have been shown to produce all types of cancer. Modes of transmission for these viruses are variable. Although there is evidence that simian virus 40 (SV40) integrates its DNA into the host-cell chromosome, a general view of how tumor viruses cause host-cell transformation has not yet been established.

25. The answer is A. *(Davis, ed 2. pp 1298-1299.)* The coxsackieviruses (groups A and B) produce a variety of illnesses, including aseptic meningitis, acute upper respiratory disease, and a paralytic disease simulating poliomyelitis. Twice the normal incidence of congenital heart lesions is found in infants whose mothers had coxsackievirus infections during the first trimester of pregnancy. Epidemic pleurodynia (Bornholm's disease, epidemic myalgia) is a coxsackieviral disease causing paroxysmal chest pain and increasing fever and debility.

26. The answer is A. *(Jawetz, ed 13. p 459.)* All patients with exanthem subitum (roseola infantum) recover promptly and without any therapy. High fever usually persists for three to four days; as the fever abates, pink maculopapules appear on the chest and trunk, where they remain for one to three days. Except for rare febrile convulsions, exanthem subitum resolves without complications.

27. The answer is D. *(Jawetz, ed 13. p 459.)* Common skin warts (verrucae) are caused by human wart virus (papilloma virus), which belongs to the papovavirus group (*pa*pilloma, *po*lyoma, and *va*cuolating viruses). Human wart viruses do not grow in culture or in laboratory animals. The virus can be spread by direct or indirect contact between individuals, by autoinoculation, or by scratching.

28. The answer is D. *(Jawetz, ed 13. p 409.)* Influenza virus is relatively stable and can be stored for one week at 4°C. Infectivity is best preserved at −70°C. Although influenza virus is sensitive to heat, treatment with a 1M concentration of magnesium sulfate ($MgSO_4$) appears to stabilize the virus against heat damage. Infectivity is destroyed by agents, such as ether and formalin, that denature proteins.

29. The answer is B. *(Jawetz, ed 13. p 366.)* Yellow fever is an arbovirus infection transmitted by the bite of the *Aedes* mosquito. A disease of South America and Africa, yellow fever is characterized by a high fever, jaundice, proteinuria, and the vomiting of black vomitus. Dengue, also caused by an *Aedes*-born arbovirus, is usually a more benign infection than yellow fever and is characterized by fever, muscle and joint pain, and lymphadenopathy. Equine encephalitis is caused by yet another arbovirus, kala-azar by a parasite, and scrub typhus by rickettsiae.

30. The answer is E. *(Jawetz, ed 13. pp 293-295.)* Poliomyelitis is diagnosed by isolating the virus, demonstrating a rising antibody titer, or both. Skin tests for polio are not in existence, although other viral conditions, such as herpes and cat scratch fever, can be uncovered by antigenic skin testing. Skin tests also can help in the diagnosis of lymphogranuloma venereum, a chlamydial disease, and sarcoidosis, the etiology of which is as yet unestablished.

31. The answer is B. *(Volk, p 471.)* Cytomegalovirus (CMV) infection is associated with a wide array of congenital disorders. The source of fetal CMV disease often is a clinically inapparent infection in the mother. The consequences of congenital CMV disease can include prematurity, hepatosplenomegaly, pneumonitis, microcephaly, mental retardation, and intrauterine or neonatal death. Infected infants can excrete the virus in their urine or saliva for long periods of time.

32. The answer is C. *(Davis, ed 2. pp 1217-1218, 1261, 1264-1268.)* Smallpox virus infection is overwhelmingly virulent and acute. The incubation period for the disease is 12 days, and the symptoms include fever, myalgia, abdominal pain, and vesicular and pustular rashes. The other viruses mentioned in the question can be harbored in a host for years without causing clinically apparent disease.

33. The answer is B. *(Davis, ed 2. pp 1212-1213.)* All the organisms listed in the question, except poliovirus, cause infection primarily at the entry organ. They may, with time, cause erosion into blood or lymphatic vessels and consequently spread to distant sites; however, this dissemination is not essential in producing the characteristic disease syndromes. Poliovirus infection, on the other hand, generally begins in the oropharynx and distal intestinal tract; dissemination occurs by way of intestinal lymphatic channels and thence into the bloodstream.

34. The answer is D. *(Jawetz, ed 13. p 402.)* Aseptic meningitis is characterized by a pleocytosis of mononuclear cells in the cerebrospinal fluid; polymorphonuclear cells predominate during the first 24 hours, but a shift to lymphocytes occurs thereafter. The cerebrospinal fluid of affected individuals is free of culturable bacteria and contains normal glucose and slightly elevated protein levels. Peripheral white blood cell counts usually are normal. Although viruses are the most common cause of aseptic meningitis, spirochetes, chlamydiae, and other microorganisms also can produce the disease.

35. The answer is B. *(Jawetz, ed 13. p 392.)* A recent report suggests that a hitherto-unidentified virus—non-A, non-B (NANB) hepatitis virus—accounts for over 90 percent of transfusion-induced hepatitis. This virus does not appear to be related immunologically to either hepatitis A or B virus. Routine tests for NANB virus do not yet exist, although recent reports have described a radioimmunoassay procedure able to detect this virus.

36. The answer is A. *(Jawetz, ed 13. pp 397, 401. Volk, p 557.)* Since 1960, the yearly incidence of rabies in the United States has remained at about one to three cases. Worldwide, however, 1,000 fatal human cases, or more, are reported annually. Dogs and cats are the most common sources of human exposure to rabies in most countries of the world; in the United States, however, wild animals—particularly bats, skunks, and foxes—are the chief offenders, accounting for 70 percent of the 3,123 cases of animal rabies reported in 1974.

37. The answer is A. *(Jawetz, ed 13. p 487.)* The simian B herpesvirus produces a fortunately rare but extremely virulent infection in humans. It causes acute encephalitis and myelitis and necrosis of the spleen; death usually results within 10 days of onset. In monkeys, simian B virus produces a latent infection associated with significant oncogenic potential. Simian virus 40 (SV40) is an oncogenic papovavirus.

38. The answer is A. *(Davis, ed 2. pp 1246-1248.)* Varicella-zoster virus, a member of the herpesvirus group, causes a usually mild, self-limited illness in children. Recurrent disease in adults who possess circulating antibody against varicella-zoster virus may be more severe and cause an inflammatory reaction in one or more of the sensory ganglia of spinal or cranial nerves. This disease, shingles, appears to result from the reactivation by trauma or other stimuli of latent varicella-zoster virus.

39. The answer is D. *(Volk, pp 528-529.)* Mild disease resulting from infection with adenovirus, respiratory syncytial virus, influenza virus, or parainfluenza virus may be indistinguishable from the common cold. Nevertheless, the entity recognized as the common cold appears to be caused most often by rhinoviruses, which are RNA, ether-stable viruses. In the number of days of work and school lost yearly, the common cold exerts an enormous socioeconomic influence in the United States.

40. The answer is B. *(Davis, ed 2. pp 1216, 1356-1357.)* Rubella (German measles) is a mild viral disease characterized by lymphadenopathy, fever, and a rash. However, approximately 30 percent of women who have clinical rubella during the first trimester of pregnancy give birth to babies who have structural abnormalities. An effective vaccine against rubella is available; however, women should **not** be vaccinated once pregnant or if intending to become pregnant within a few months of inoculation.

41. The answer is E. *(Jawetz, ed 13. p 370.)* Lassa fever first was recognized in Nigeria in 1969. The causative arenavirus, an RNA virus that cross-reacts immunologically with lymphocytic choriomeningitis (LCM) virus, is extremely virulent. The disease is characterized initially by high fever and pharyngitis; its course may be followed by serial determination of immunofluorescent antibody titers.

42. The answer is E. *(Davis, ed 2. p 1348.)* Administration of live attenuated measles virus vaccine is the most effective control measure against rubeola. It induces an antibody response in almost 100 percent of inoculated children; however, it also produces a rash or fever (or both) in 10 to 25 percent of these children. Measles vaccination does not have neurologic contraindications. Effective immunity may last up to eight years; a complete evaluation of the vaccine's period of protection is not yet possible.

43. The answer is C. *(Davis, ed 2. pp 1410-1413. Jawetz, ed 13. pp 386-388.)* Hepatitis-associated surface antigen (HB_sAg) appears in the serum late in the incubation period of hepatitis B (30 to 50 days after infection) and may persist long after symptoms subside. HB_sAg contains protein but no nucleic acids. It has been identified as the outer-envelope component of Dane particles, which contain an inner core of DNA and are surmised to be the viral agent of hepatitis B.

44. The answer is A. *(Volk, p 547.)* The testes and ovaries are involved fairly commonly in individuals who have mumps. Of male individuals who develop mumps after the age of 13 years, 20 to 30 percent will have orchitis. The mumps incubation period is usually about 20 days, and parotiditis, the major clinical feature, lasts about one week. Mumps virus (one of the paramyxoviruses) rarely, if ever, affects the urinary tract, bones, skin, or gastrointestinal system.

45. The answer is C. *(Jawetz, ed 13. p 160.)* In fully susceptible individuals, no reaction to smallpox vaccination is seen until at least day three or four, when a papule surrounded by a narrow areola appears. The size of the papule increases until vesiculation occurs on day five or six. Maximum vesicular size is reached by the seventh to ninth day, after which time the lesion becomes pustular. Desiccation and regression follow, leaving a depressed pink scar. Repeat vaccination is indicated if this reaction is not produced.

46. The answer is C. *(Jawetz, ed 13. p 426.)* Maternal rubella infection during the first month of pregnancy produces abnormalities in approximately 80 percent of offspring; infection during the third month of gestation produces anomalies in only 15 percent of offspring. Rubella virus does not actually destroy fetal cells but rather slows the growth rate of infected cells. Among the defects associated with congenital rubella syndrome are patent ductus arteriosus, atrial and ventricular septal defects, and other anomalies of the heart and great vessels; cataracts, chorioretinitis, and other lesions of the eyes; deafness; microcephaly; hepatosplenomegaly; and mental retardation. Affected infants commonly have elevated levels of immunoglobulin M (IgM) and low levels of IgG and IgA.

47. The answer is E. *(Jawetz, ed 13. p 354.)* The definitive diagnosis of rabies in humans is based on the finding of Negri bodies, which are cytoplasmic inclusions in the nerve cells of the spinal cord and brain, especially in the hippocampus. Negri bodies are eosinophilic and generally spherical in shape; several may appear in a given cell. Negri bodies, though pathognomonic for rabies, are not found in all cases of the disease.

48. The answer is E. *(Jawetz, ed 13. p 374.)* Poliomyelitis, a viral disease that can cause flaccid paralysis through the destruction of lower motor neurons, is spread primarily by the fecal-oral route. No evidence exists for significant spread of the virus by either airborne or venereal transmission. Although polio can be controlled by vaccination, the disease has not been eradicated; this fact must be stressed to parents, some of whom have adopted a dangerous nonchalance toward the need to vaccinate their children against polio.

49. The answer is D. *(Jawetz, ed 13. p 436.)* Routine vaccination of infants and children for smallpox has been discontinued in the United States, both because the risk of contracting the disease is so low and because the complications of smallpox vaccination, including generalized vaccinia eruption, postvaccine encephalitis, and fetal vaccinia, are significant. Because eradication of smallpox worldwide by the World Health Organization has been very effective, U.S. citizens traveling abroad no longer require vaccination. Pregnancy, immune deficiencies, and eczema and other chronic dermatitides are contraindications against smallpox vaccination.

50. The answer is E. *(Volk, p 579.)* The specimen most likely to be positive for herpes simplex in the woman described in the question would be fluid expressed from the vesicular lesions. Herpesvirus hominis type 1 (herpes simplex) also has been isolated from cells of blood, urine, and throat washings. Other disorders caused by this virus include keratoconjunctivitis, meningoencephalitis, and cold sores.

51. The answer is D. *(Jawetz, ed 13. pp 453-454.)* Infectious mononucleosis characteristically is accompanied by splenomegaly, the appearance of unique sheep-cell hemagglutinins, an elevated peripheral white blood cell count, and the presence of atypical lymphocytes known as Downey cells. Patients also may develop antibodies to the causative Epstein-Barr (EB) virus as measured by immunofluorescent staining of virus-bearing cells.

52. The answer is B. *(Jawetz, ed 13. pp 408-415.)* Type A influenza virus infections are more serious than those caused by either type B or type C viruses. Transmission is person-to-person; however, there are no known human carriers, making interepidemic maintenance of the disease a puzzle. Influenza has a fairly low case-fatality rate, but in the pandemic form it does cause a significant number of deaths.

53. The answer is A. *(Volk, p 506.)* The *e* antigen seems to be related to the Dane particle, which is presumed to be the intact hepatitis B virus. Possession of the *e* antigen suggests active disease and, thus, an increased risk of transmission of hepatitis to others. Ingestion of raw shellfish, notably clams and oysters, has been associated with cases of hepatitis A.

54. The answer is E. *(Volk, p 528.)* The majority of common colds in adults are caused by rhinoviruses. The other disorders listed in the question also are associated with infection by rhinoviruses but occur more frequently in young children than in adults. Other viruses, including adenoviruses, reoviruses, and respiratory syncytial viruses, also can cause common colds.

55. The answer is A. *(Davis, ed 2. pp 1248, 1267.)* The eosinophilic inclusion bodies are intranuclear in chickenpox and cytoplasmic in smallpox (Guarnieri bodies). These inclusion bodies are composed of clumps of virus and viral antigens. Varicella characteristically is associated with the presence of multinuclear giant cells.

56. The answer is E (all). *(Volk, pp 142-143.)* Robert Koch outlined conditions needed to establish that a particular microorganism bears an etiologic relationship to a given disease. According to Koch's "postulates," an investigator seeking to prove this relationship must (1) demonstrate that the organism is associated regularly with the disease, (2) isolate and grow the organism in pure culture, (3) produce the disease in an animal by inoculating it with the organism, and (4) isolate the organism from the diseased animal. For his work on the use of tuberculin to test for tuberculosis, Robert Koch was awarded the 1905 Nobel Prize in Physiology and Medicine.

57. The answer is D (4). *(Davis, ed 2. pp 1010, 1023-1025.)* In addition to their nucleic acid core and protein coat, many groups of viruses contain an outer lipid envelope. Although all viral genomes are comprised of a single type of nucleic acid (either RNA or DNA), the nucleic acid structure is not always single-stranded; poxviruses and herpesviruses, for example, contain double-stranded DNA. Many viruses carry specific enzymes, frequently ones involved in nucleic acid metabolism. Tumor viruses may package cellular RNA.

58. The answer is E (all). *(Jawetz, ed 13. pp 320-322.)* Interferon is a protein that alters cell metabolism to inhibit viral replication. It induces the formation of a second protein, which interferes with the translation of viral messenger RNA. Production of interferon has been demonstrated when cells in tissue culture are challenged with viruses, rickettsiae, endotoxin, or synthetic double-stranded polynucleotides. Interferon confers species-specific, not virus-specific, protection for cells.

59. The answer is B (1, 3). *(Jawetz, ed 13. pp 302-303.)* The virion, which is the complete infective virus particle, is identical with the capsid and its nucleic acid core in adenoviruses and picornaviruses. In more complex virions, such as herpesviruses, poxviruses, and myxoviruses, the virion consists of the nucleocapsid plus a surrounding envelope or coat. Viral envelopes can contain protein, lipid, and carbohydrate.

60. The answer is A (1, 2, 3). *(Davis, ed 2. p 1207.)* Viral cytopathic effects are thought to include a change in the host cell's macromolecular synthesis and cell-membrane structure. Viruses may produce cytopathic changes without forming infectious virions and without replicating infectious virus. A particular cytopathic effect is not necessarily associated with a specific virus.

61. The answer is A (1, 2, 3). *(Jawetz, ed 13. pp 385, 397, 419, 422.)* Rabies virus is inactivated in one hour at 50°C and in five minutes at 60°C. Mumps virus and measles virus are destroyed at 56°C after twenty minutes and one hour, respectively. Hepatitis A virus is resistant to moderate heat and chemical agents; its infectivity is destroyed in five minutes at 100°C.

62. The answer is A (1, 2, 3). *(Davis, ed 2. p 1190.)* A humoral antibody response can be elicited by viral products as well as by complete virions. Moreover, cellular antibodies can react with internal as well as surface components of a virion. A cellular immune response also can be elicited by virus-specific proteins on the infected cell's surface membrane.

63. The answer is C (2, 4). *(Davis, ed 2. pp 1207-1208.)* Viral inclusion bodies may result from either the localized synthesis of virion subcomponents or the storage of finished particles. They are often of crucial diagnostic importance, e.g., in distinguishing between smallpox and chickenpox. Viral inclusion bodies, which can be found in a cell's nucleus or cytoplasm (or both), disrupt or kill the host cell.

64. The answer is E (all). *(Davis, ed 2. p 1206.)* Successfully virulent viruses are able to multiply at temperatures found in a febrile host and overcome host-cell resistance factors. Among these factors is host-cell production of interferon, which can inhibit viral replication. The virulence of viruses is under polygenic control.

65. The answer is E (all). *(Davis, ed 2. p 1191.)* Virus-antibody complexes can be dissociated by changing pH, by sonic vibration, or by competition from inactivated virions. Neutralization is a function of the reaction between host cell, virion, and antibody. Identical virion-antibody mixtures display divergent infectivity levels when assayed in different host cells. Virion-antibody complexes that initially are reversible become irreversible with time.

66. The answer is C (2, 4). *(Davis, ed 2. pp 1173-1181.)* Interferon is postulated to have a wide spectrum of antiviral activity. Its potential as an antiviral agent seems to be limited, however, by its reversible action, short period of effectiveness, and inability to protect already infected cells. Interferon inhibits the formation of complexes between viral messenger RNA and host-cell ribosomes.

67. The answer is E (all). *(Davis, ed 2. pp 1293, 1375, 1413-1414.)* Both forms of viral hepatitis are responsive to gamma globulin therapy. Hyperimmune rabies antiserum prolongs the incubation period of the disease, allowing the patient more time to mount an immune response to the vaccine. Although not a primary form of treatment for individuals who have poliomyelitis, passive immunization with pooled gamma globulin can offer adequate protection against the disease.

68. The answer is E (all). *(Jawetz, ed 13. pp 453-454.)* Epstein-Barr (EB) virus, a member of the herpesvirus group, has been associated with infectious mononucleosis, nasopharyngeal carcinoma, and Burkitt's lymphoma. It is not yet clear whether EB virus is the etiologic agent of the malignancies or merely a passenger virus. Herpesvirus type 2 antibodies have been found in association with carcinoma of the cervix, but again it is not known whether this association is etiologic. In at least one cervical cancer biopsy, herpesvirus DNA and messenger RNA were reported present.

69. The answer is D (4). *(Jawetz, ed 13. p 404.)* Slow viruses produce progressive neurologic disease and may have incubation periods of up to five years before their clinical manifestations become apparent. Progressive multifocal leukoencephalopathy (PML), subacute sclerosing panencephalitis (SSPE), kuru, and Creutzfeldt-Jakob disease are human diseases caused by slow viruses; other chronic diseases undoubtedly will someday prove to be of similar origin. SSPE apparently is caused by a virus closely resembling the virus causing measles, and affected individuals might possess antibodies against the measles virus. Scrapie is one of several slow-virus diseases of animals.

70. The answer is E (all). *(Jawetz, ed 13. p 327.)* Viruses may be transmitted by direct person-to-person contact through droplet or aerosol infection (e.g., influenza). Rabies is transmitted through a bite wound from an infected animal. Arboviruses use arthropods as vectors. Poliomyelitis is transmitted through the alimentary tract. Several viruses have multiple means of transmission and, thus, multiple epidemiologic patterns.

71. The answer is A (1, 2, 3). *(Davis, ed 2. p 1389.)* Western equine encephalitis (WEE) is caused by a group A togavirus. The natural reservoir for the infection is wild birds, and mosquitoes appear to be the primary vectors in the transmission to humans and horses. Clinically similar diseases, including eastern equine encephalitis, St. Louis encephalitis, and Japanese B encephalitis, may be caused by a variety of related arboviruses. In general, these encephalitides are characterized by the sudden onset of headache, fever, and nausea, 4 to 21 days after inoculation.

72. The answer is B (1, 3). *(Jawetz, ed 13. pp 389-390.)* At present, several methods are available for detection of hepatitis B surface antigen (HB$_s$Ag) and antibodies to this antigen. Among these methods are complement fixation, gel diffusion, counterelectrophoresis, and radioimmunoassay. The isolation of Dane particles, which are intact virions, and the routine detection of e antigen are not yet possible.

73. The answer is D (4). *(Volk, p 470.)* The cytomegalovirus causing human disease is species-specific for human hosts. Infants are affected through intrauterine or early postnatal infection; the mechanism of transmission of the virus for the general population remains unknown. Lymphocytic choriomeningitis (LCM) virus is carried by mice, and variola (smallpox) virus and the enterovirus causing poliomyelitis can produce disease in monkeys.

74. The answer is D (4). *(Davis, ed 2. pp 1234, 1419.)* Adenovirus types 12 and 18 are able to cause tumors in neonatal hamsters, rats, and mice. In human cells, however, these viruses produce infectious adenovirus progeny — i.e., the oncogenic potential of the organisms remains unexpressed. In addition to these two types of adenovirus, coronaviruses and certain herpesviruses and papovaviruses have been shown to have oncogenic potential.

75. The answer is B (1, 3). *(Jawetz, ed 13. p 429.)* Smallpox and chickenpox are caused by DNA viruses (poxvirus and herpesvirus, respectively). Measles (both rubeola and rubella) and yellow fever are caused by RNA-containing viruses — rubeola by paramyxovirus and rubella and yellow fever by togavirus. In general, most DNA viruses have a double-stranded genome and most RNA viruses a single-stranded genome; all viral families mentioned above conform to these generalizations.

76. The answer is B (1, 3). *(Davis, ed 2. pp 1378-1381.)* Arboviruses (*ar*thropod-*bo*rne viruses) may or may not be surrounded by a lipid envelope, although most are inactivated by lipid solvents like ether, and may contain either double-stranded or single-stranded RNA. Physicochemical studies have demonstrated a great heterogeneity among these viruses. Arboviruses cause disease in vertebrates; in humans, encephalitis is a frequent arbovirus illness.

77. The answer is D (4). *(Jawetz, ed 13. pp 358-362.)* Infectious encephalitides rarely are produced by bacteria. They are practically always associated with fever and are essentially untreatable. Neurologic sequellae are common.

78. The answer is E (all). *(Jawetz, ed 13. p 381.)* Echoviruses, in addition to causing all the illnesses listed in the question, also can produce aseptic meningitis, infantile diarrhea, and vaginitis. Many of the 30 serotypes of echovirus are still orphans — i.e., not yet associated with disease. Enteroviruses produce transitory infections, primarily during the summer and autumn; the prevalence of these infections in children of poorer families is significantly higher than in other children.

79. The answer is A (1, 2, 3). *(Davis, ed 2. p 1299.)* In individuals who have epidemic pleurodynia, chest pain and fever usually occur simultaneously and abruptly. Pain is severe and aggravated by movement; abdominal muscle spasms develop in about 50 percent of affected individuals. The illness, caused by a group B coxsackievirus, is self-limited, and although recovery is complete, relapses are common. Orchitis is an infrequent complication in male individuals having a relapse of epidemic pleurodynia.

80. The answer is C (2, 4). *(Davis, ed 2. pp 1362-1366.)* Coronaviruses were discovered in 1965 during a search for the etiologic agent of the common cold. The virion is known to contain RNA, but other elements of its structure are uncertain. It is thought to be a major agent of the common cold, especially in older children and adults.

81. The answer is E (all). *(Jawetz, ed 13. p 418.)* Paramyxoviruses are a large group of RNA viruses that cause respiratory illness, measles, mumps, and neurologic disease. The presence of a hemagglutinin and a hemolysin aids in the laboratory differentiation of these viruses. Paramyxoviruses have the ability to produce persistent infection that is noncytocidal for cells in culture.

82. The answer is C (2, 4). *(Jawetz, ed 13. p 398.)* Rabies virus produces an acute neurologic infection carrying a fatality rate of almost 100 percent. Generally speaking, animals can transmit the virus for only a few days (five for dogs, for example) before their disease becomes apparent; the bat is a notable exception to this rule. Rabies virus can be isolated from the saliva of infected individuals.

83. The answer is A (1, 2, 3). *(Davis, ed 2. pp 1250-1251.)* Though cytomegalovirus infection is common, it only rarely causes clinically apparent disease. Lesions characteristic of cytomegaloviral infection are found in up to 10 percent of stillborn babies; however, cytomegalovirus, which can be transmitted transplacentally, usually is not the cause of death. Children and adults with immunosuppressive problems are susceptible to active disease. Control of cytomegalovirus infection is not yet available.

84. The answer is C (2, 4). *(Jawetz, ed 13. p 418.)* Hemadsorption activity of inoculated cultures suspected of containing paramyxovirus (e.g., mumps virus) should be tested at 4°C. At this temperature, the enzyme neuraminidase is unable to exert its inhibitory effect on hemagglutination. At 35°C, on the other hand, neuraminidase is active and destroys hemagglutinin receptors.

85. The answer is E (all). *(Jawetz, ed 13. p 449.)* Infection with varicella-zoster virus is diagnosed readily by examining cells taken from skin scrapings of an individual's vesicular lesions. Fifteen percent of adults who have varicella develop pneumonia. Encephalomyelitis is a rare complication in children. Generalized zoster, which occurs in five percent of affected patients, may lead to death in the presence of underlying disease. Varicella-zoster virus is morphologically identical to herpes simplex virus.

86. The answer is D (4). *(Volk, p 567.)* Slow-virus diseases such as kuru, sub-acute sclerosing panencephalitis, and Creutzfeldt-Jakob disease have not yielded to conventional techniques of viral isolation or detection. Although no etiologic agent has been described for kuru, the disease described in the question, injection of suspensions of human brains into nonhuman primates has resulted in disease. Such an observation suggests an infectious etiology for kuru.

87. The answer is E (all). *(Jawetz, ed 13. p 453.)* Fifty to eighty percent of patients who have infectious mononucleosis develop an increased titer of heterophil antibodies that can be demonstrated by agglutination of sheep erythrocytes. Affected individuals have lymphadenopathy, and the presence of atypical lymphocytes in the peripheral smear is characteristic. The Epstein-Barr virus is thought to be the etiologic agent.

88. The answer is D (4). *(Jawetz, ed 13. pp 385-396.)* Hepatitis A virus (HAV) disease, formerly known as "infectious hepatitis," is a worldwide public-health problem. Although its mode of transmission classically has been accepted to be the fecal-oral route, hepatitis A also may be transmitted parenterally, the route more commonly associated with spread of hepatitis B ("serum hepatitis"). Viral hepatitis type A has an incubation period of 15 to 45 days (average, 25 to 30); for type B, on the other hand, it is 50 to 180 days (average, 60 to 90). These forms of viral hepatitis can be hard to distinguish clinically.

89. The answer is A (1, 2, 3). *(Jawetz, ed 13. p 421.)* Parainfluenza viruses form a subgroup of paramyxoviruses and are recognized by their ability to lyse, as well as agglutinate, erythrocytes. Parainfluenza virus types 1 and 2 appear to be the main agents causing croup in children; antigenically, type 2 is unrelated to all other paramyxoviruses except mumps virus. Parainfluenza virus type 4 is not known to cause any human illness.

90. The answer is E (all). *(Jawetz, ed 13. p 379.)* Epidemic pleurodynia, which is characterized by paroxysmal chest pain, and herpangina, which is marked by the acute onset of fever, dysphagia, vesiculopapular eruptions, and other symptoms, are caused by coxsackievirus infection. Coxsackieviruses also can cause aseptic meningitis, hand-foot-and-mouth disease, and colds. They are becoming increasingly well known as a primary cause of myocarditis in both adults and children. Neonatal disease due to coxsackievirus group B is characterized by lethargy, fever, vomiting, and, in some cases, initial manifestations of anorexia and diarrhea.

91. The answer is C (2, 4). *(Jawetz, ed 13. p 431.)* Guarnieri bodies are cytoplasmic inclusions characteristically found in cells infected with variola major (smallpox). Variola virus causes a transient viremia that is followed by infection of reticuloendothelial cells throughout the body. A secondary phase of multiplication within reticuloendothelial cells leads to a second, more intense viremia and clinical disease. The smallpox lesions in a given area tend to be in the same stage of histologic development.

92. The answer is A (1, 2, 3). *(Davis, ed 2. pp 1346-1347.)* Measles virus has been isolated from patients with giant cell pneumonia. This virus also has been implicated, on the basis of high antibody titers and other data, in the pathogenesis of the slow-viral infection subacute sclerosing panencephalitis (SSPE). The incidence of encephalomyelitis associated with measles is approximately 1 in 10,000 cases. The presence of red Koplik's spots on the buccal mucosa is a characteristic of measles.

93. The answer is B (1, 3). *(Davis, ed 2. p 1275.)* Molluscum contagiosum virus causes a rare skin disease primarily affecting children and young adults. The virus stimulates cell division in uninfected neighboring cells, even though DNA synthesis ceases in infected cells. The chronic lesion is restricted to the epithelium of the skin of the back, face, arms, legs, and genitals. Electron microscopy shows virions that are identical to other poxviruses.

94. The answer is B (1, 3). *(Davis, ed 2. pp 1387-1388.)* Dengue (breakbone fever) is caused by a group B togavirus that is transmitted by mosquitoes. The clinical syndrome usually consists of a mild systemic disease characterized by severe joint and muscle pain, headache, fever, lymphadenopathy, and a maculopapular rash. In some epidemics, hemorrhagic dengue, a more severe syndrome, may be prominent; shock and, occasionally, death result.

95. The answer is D (4). *(Jawetz, ed 13. p 447.)* Acute herpetic gingivostomatitis (Vincent's stomatitis) is the most common clinical disease due to primary infection with type 1 herpesvirus hominis. The disorder is characterized by fever, local lymphadenopathy, and extensive lesions of the mucous membranes of the oral cavity. Vincent's stomatitis occurs most frequently in small children.

96. The answer is E (all). *(Volk, p 532.)* Rotavirus is a new viral entity thought to be a major cause of acute diarrhea in newborn infants. Three-quarters of all adults have antibodies against rotavirus; passive transfer of these antibodies to the baby, especially through the colostrum, seems to be protective. Although vaccination would be expected to be of little use to the neonate, it might effectively immunize pregnant mothers.

97. The answer is D (4). *(Jawetz, ed 13. p 419.)* Much of the public's understanding of mumps is based on old wives' tales that are without any scientific basis. For example, natural mumps infection confers immunity after a single infection, even if the infection was a unilateral, not bilateral, parotiditis. Also, sterility from mumps orchitis is rare; only 20 percent of male individuals older than 13 years of age develop orchitis. Lastly, the virus is maintained exclusively in human populations; canine reservoirs are not known. The mumps vaccine is a live-attenuated-virus vaccine derived from chick-embryo tissue culture.

98. The answer is D (4). *(Jawetz, ed 13. pp 451-452.)* Clinical manifestations of cytomegalovirus (CMV) infection may not be readily apparent at birth. Thus, in a newborn infant with a 1:32 titer to CMV, it is necessary to determine whether the antibodies were passed transplacentally from the mother (these antibodies would be immunoglobulin G) or produced by the fetus in response to an in-utero infection (IgM). A newborn infant who is infected excretes large numbers of virus particles in the urine and, therefore, places other neonates at risk for contracting CMV disease.

99. The answer is B (1, 3). *(Davis, ed 2. pp 1315-1317.)* Influenza A viruses have undergone three major antigenic shifts since the 1933 pandemic, and each shift is accompanied by a high degree of susceptibility in a large part of the population. This antigenic variation, or drift, creates serious problems in developing effective vaccines. Minor antigenic changes, occurring every two to three years, have had less impact, due to cross-reactivity of the neutralizing antibodies. Variations in influenza B viruses have been less marked and less frequent and have all been subject to immune cross-reactivity.

100. The answer is B (1, 3). *(Jawetz, ed 13. pp 360-361.)* St. Louis encephalitis, which is caused by a togavirus, is characterized by marked degeneration of neurons, especially in the midbrain and medulla. The reservoir of St. Louis encephalitis virus is wild birds, and the disease is transmitted to humans by mosquitoes. St. Louis encephalitis occurs six times as frequently in adults as in children; in the 1966 epidemic, no one less than the age of 45 years was reported to have died from the disease. The incubation period of St. Louis encephalitis is one to three weeks.

101. The answer is A (1, 2, 3). *(Jawetz, ed 13. pp 397-402.)* Rabies, an extremely virulent disease in humans, is caused by an ether-sensitive RNA virus (a rhabdovirus) that forms Negri bodies in the cytoplasm of infected nerve cells. Chiropteran, or bat, rabies, which may differ from canine strains, can produce an ascending myelitis and spreading paralysis resembling acute ascending paralysis (Landry's paralysis) or the Guillain-Barré syndrome. It is interesting to note that this type of paralysis has been reported to follow antirabies inoculation. The incubation period of human rabies is two weeks to a year.

102. The answer is B (1, 3). *(Davis, ed 2. pp 1242-1245.)* Initial herpes simplex virus infection is often inapparent and occurs through a break in the skin or mucous membranes, such as in the eye, throat, or genitals. Latent infection often persists at the initial site despite high antibody titers. Recurrent disease can be triggered by temperature change, emotional distress, and hormonal factors. Type 1 herpes simplex virus is usually, but not exclusively, associated with ocular and oral lesions; type 2 is usually, but not exclusively, associated with genital and anal lesions.

103. The answer is E (all). *(Davis, ed 2. p 1216.)* Several viruses are known to be teratogenic—i.e., to produce congenital malformations—in humans. Rubella infection during pregnancy is associated with congenital fetal anomalies, and rubella remains the major viral cause of fetal death. Cytomegalovirus may cause microcephaly, herpesvirus may lead to central nervous system disease, and coxsackievirus infection may produce cardiac lesions.

104. The answer is D (4). *(Briody, pp 599-601.)* All of Koch's postulates have been verified for the relationship between infectious mononucleosis and Epstein-Barr virus, a type of herpesvirus. However, the relationship between this virus and Burkitt's lymphoma, sarcoid, systemic lupus erythematosus (SLE), and leprosy is less clear. Infectious mononucleosis is most common in young adults (15 to 20 years of age) and is very rare in young children. There is no specific treatment. Heterophil antibody titer is helpful in diagnosis.

105. The answer is A (1, 2, 3). *(Davis, ed 2. p 1274.)* N-methylisatin-β-thio-semicarbazone (methisazone) inhibits viral assembly and thus theoretically would be effective in treating persons who have smallpox. So, too, would vaccinia hyperimmune serum and transfer factor, which bolster most defense mechanisms. Tetracycline, however, would be of no benefit in the treatment of these persons.

106-110. The answers are: 106-C, 107-D, 108-D, 109-B, 110-A. *(Jawetz, ed 13. p 330.)* Immunization against smallpox entails the use of a vaccine containing live active virus. The sources of smallpox vaccine are calf or sheep lymph and chorioallantois. Routine immunization of U.S. citizens against smallpox is no longer practiced.

In developed countries, it is recommended that the general population be immunized routinely against measles, rubella, mumps, and poliomyelitis. Live-attenuated-virus vaccines for these diseases are derived from a variety of tissue cultures; mumps and measles vaccines, for instance, are made from chick embryo tissue culture, and poliomyelitis vaccine from tissue culture of monkey kidney cells. Live-attenuated-virus vaccines are associated with a longer-lasting antibody response than other types of vaccine; however, there is a risk that the virus can revert to a more virulent form once in the body.

Under the appropriate circumstances, inactivated-virus vaccines are given for rabies and eastern equine encephalitis. Rabies vaccine is made from duck embryo treated with ultraviolet light or phenol; vaccine for eastern equine encephalitis is derived from chick embryo cell culture. Vaccines containing inactive virus confer a briefer immunity than those with live virus.

111-114. The answers are: 111-A, 112-B, 113-E, 114-C. *(Jawetz, ed 13. pp 385-389.)* It is difficult to make a reliable clinical distinction between disease caused by hepatitis A virus (HAV) or hepatitis B virus (HBV). Hepatitis A usually is contracted by the fecal-oral route, whereas hepatitis B typically is transmitted parenterally. Recent studies, however, indicate that nonparenteral transfer of HBV is possible.

The presence of hepatitis surface antigen ($HB_s Ag$) is associated closely with hepatitis B infection. This antigen can be detected by radioimmunoassay.

The core antigen of hepatitis B virus—$HB_c Ag$—constitutes the inner component of the intact virus particle. Both a DNA template (double-stranded DNA) and DNA polymerase have been noted to be associated with core antigen. Its clinical significance is not understood.

Bacteriology

DIRECTIONS: Each question below contains five suggested answers. Choose the **one best** response to each question.

115. Cell envelopes of both gram-positive and gram-negative bacteria are composed of complex macromolecules. Which of the following statements describes both gram-positive and gram-negative cell envelopes?

(A) They contain a significant amount of teichoic acids
(B) Their antigenic specificity is determined by the polysaccharide O antigen
(C) They act as a barrier to the extraction of crystal violet-iodine by alcohol
(D) They contain the common amino acids
(E) They are a diffusion barrier to large charged molecules

116. Complexes of complement and specific antibody are bactericidal to certain bacteria. The lytic action of complement would be most effective against

(A) Group B *Streptococcus*
(B) *Corynebacterium diphtheriae*
(C) *Klebsiella pneumoniae*
(D) *Mycobacterium tuberculosis*
(E) *Clostridium tetani*

117. A technologist notes that both the quality-control strain of *Staphylococcus aureus* and a strain of *Escherichia coli* grow and produce a yellow color on mannitol salt agar, a selective medium for staphylococci. The technologist should conclude that the agar

(A) contains no mannitol
(B) has a reduced salt content
(C) is most likely a 3% solution
(D) is outdated
(E) had been contaminated with *E. coli* during its manufacture

118. Which of the following tests would NOT be useful in making a diagnosis of syphilis?

(A) Frei test
(B) Fluorescent treponemal antibody-absorption test
(C) Venereal Disease Research Laboratories (VDRL) test
(D) Automated reagin test
(E) Wassermann test

119. The Dick test is a skin test used in the diagnosis of

(A) lymphogranuloma venereum
(B) scarlet fever
(C) tuberculosis
(D) sarcoidosis *Kveim*
(E) histoplasmosis

120. Cerebrospinal fluid drawn from a newborn baby is noted on Gram stain to contain gram-positive cocci; the following morning, beta-hemolytic colonies resembling streptococci are observed on blood agar plates. Steps in the identification of these streptococci should include

(A) hippurate hydrolysis
(B) bacitracin susceptibility
(C) optochin sensitivity
(D) coagulase testing
(E) oxidase testing

121. Assuming that the average achievable serum level of gentamicin is 6 to 8 micrograms per ml (μg/ml), which of the following bacteria is susceptible to gentamicin?

(A) *Escherichia coli* with a minimum inhibitory concentration (MIC) of 10 μg/ml
(B) *E. coli* with a MIC of 12 μg/ml
(C) *Klebsiella* with a MIC of 0.25 μg/ml
(D) *Klebsiella* with a MIC of 6.0 μg/ml
(E) *Klebsiella* with a MIC of 20 μg/ml

122. Which of the following bacteria ferments pyruvate into butylene glycol?

(A) *Streptococcus*
(B) *Neisseria*
(C) *Enterobacter*
(D) *Clostridium*
(E) *Shigella*

123. An individual comes to an emergency room because of an infected dog bite. The wound is found to contain small gram-negative rods. The most likely cause of infection is

(A) *Escherichia coli*
(B) *Haemophilus influenzae*
(C) *Pasteurella multocida*
(D) *Brucella canis*
(E) *Klebsiella rhinoscleromatis*

124. Leprosy is still an international public-health problem. The drug of choice for treatment of *Mycobacterium leprae* infection is

(A) cycloserine
(B) dapsone
(C) penicillin
(D) tetracycline
(E) streptomycin

125. A woman develops fever, rash, and polyarthralgia during menstruation. Gram stain of the synovial fluid of her right knee reveals gram-negative diplococci. Which of the following organisms is most likely to be responsible?

(A) *Haemophilus influenzae*
(B) *Staphylococcus aureus*
(C) *Neisseria gonorrhoeae*
(D) *Pseudomonas aeruginosa*
(E) *Mycobacterium tuberculosis*

126. Löffler's medium is used primarily to culture

(A) *Actinomyces israelii*
(B) *Corynebacterium diphtheriae*
(C) *Neisseria meningitidis*
(D) *Neisseria gonorrhoeae*
(E) *Salmonella typhi*

127. The quellung test is used for the direct identification of capsule-containing bacteria. This test is based on the

(A) absorption of water by the hygroscopic capsular polysaccharide
(B) increased visualization of the capsule following the addition of specific antisera
(C) increased staining of the capsule when methylene blue is added
(D) specific staining of the capsular polysaccharide with Sudan black
(E) appearance of a red color in the medium following addition of quellung reagent

128. In the Schick test, diphtheria toxin is inoculated into an individual's right arm and toxoid into the left. An individual having no reaction in either arm is demonstrating

(A) immunity to diphtheria and hypersensitivity to extraneous materials in the injections
(B) susceptibility to diphtheria and hypersensitivity
(C) immunity to diphtheria and no hypersensitivity
(D) susceptibility to diphtheria and no hypersensitivity
(E) nothing conclusive

129. During a three-week period, several babies have become infected with *Acinetobacter*. The only common finding in the histories of their illnesses is the use of a mist tent. The most likely source of the organism would be

(A) dust on the plastic surface of the mist tent
(B) the oxygen supply for the mist tent
(C) the water supply for the mist tent
(D) the grease on the mist-tent fittings
(E) dust on the floor

130. The Schultz-Carlton reaction involves erythrogenic toxin and its antibody. This test is specific for

(A) the presence of streptococcal infection
(B) the presence of scarlet fever
(C) susceptibility to streptococcal infection
(D) susceptibility to scarlet fever
(E) immunity to streptococcal infection

131. A man who has a penile chancre appears in a hospital's emergency room. The VDRL test is negative. The most appropriate course of action for the physician in charge would be to

(A) send the patient home untreated
(B) repeat the VDRL test in 10 hours
(C) perform darkfield microscopy for treponemes
(D) swab the chancre and culture on Thayer-Martin agar
(E) Gram-stain the chancre fluid

132. The serologic grouping of streptococci is a function of which of the following cellular components?

(A) C-carbohydrate
(B) M-protein
(C) T-protein
(D) Hyaluronic acid
(E) Teichoic acid

133. Granulomatous lesions, which are circumscribed nodular reactions to irritating stimuli, are associated with all of the following diseases EXCEPT

(A) cat-scratch fever
(B) coccidioidomycosis
(C) tuberculosis
(D) sarcoidosis
(E) salmonellosis

134. Which of the following disorders is NOT associated with the effects of an exotoxin?

(A) Tetanus
(B) Botulism
(C) *Shigella* dysentery
(D) Diphtheria
(E) Disseminated intravascular coagulation

135. Fever of unknown origin in a farmer who raises goats would most likely be caused by which of the following organisms?

(A) *Brucella melitensis*
(B) *Clostridium novyi*
(C) *Treponema pallidum*
(D) *Histoplasma capsulatum*
(E) *Mycobacterium tuberculosis*

136. The fermentation pattern for three strains of gram-negative cocci is given below. (Only strain C grows on plain nutrient agar.) The data indicate that strain A is

	Acid produced from		
	Maltose	Dextrose	Sucrose
Strain **A**	+	+	−
Strain **B**	−	+	−
Strain **C**	−	−	−

(A) *Branhamella catarrhalis*
(B) *Neisseria flavescens*
(C) *N. gonorrhoeae*
(D) *N. meningitidis*
(E) *N. sicca*

---

137. Cholera is a toxigenic dysenteric disease common in many parts of the world. In the treatment of individuals who have cholera, the use of a drug that inhibits adenyl cyclase would be expected to

(A) kill the patient immediately
(B) eradicate the organism
(C) increase fluid secretion
(D) potentiate the action of cholera toxin
(E) block the action of cholera toxin

138. Which of the following statements about subacute bacterial endocarditis is true?

(A) It usually arises in an undamaged endocardium
(B) It is rapidly progressive
(C) It is caused most often by β-hemolytic streptococci
(D) It is caused most often by non-hemolytic streptococci
(E) It often arises as a complication of dental manipulation

139. Shigellae may be distinguished from salmonellae by their

(A) lack of motility
(B) urease production
(C) positive methyl-red test
(D) positive mannitol-fermentation test
(E) negative Voges-Proskauer reaction

140. A 68-year-old alcoholic man experiences the sudden onset of fever, shaking chills, sharp pleuritic pain, and production of "rusty" sputum. He probably has an infection caused by

(A) *Haemophilus influenzae*
(B) *Streptococcus pneumoniae*
(C) *Mycoplasma pneumoniae*
(D) *Eikenella corrodens*
(E) *Neisseria gonorrhoeae*

141. The infection of burns and wounds often is associated with

(A) *Pseudomonas*
(B) *Salmonella*
(C) *Haemophilus*
(D) *Mycobacterium*
(E) *Mycoplasma*

142. The most common site of *Escherichia coli* infection is the

(A) gallbladder
(B) gastrointestinal tract
(C) peritoneum
(D) respiratory tract
(E) urinary tract

143. *Staphylococcus aureus* can produce a severe food poisoning that results from the ingestion of

(A) enterotoxin
(B) hemolysin
(C) leukocidin
(D) coagulase
(E) penicillinase

144. The organisms most likely to exhibit metachromatic Babès-Ernst bodies when stained with aniline dyes would be

(A) neisseriae
(B) leptospirae
(C) corynebacteria
(D) pneumococci
(E) serratiae

145. Recurrent episodes of which of the following infections commonly are associated with individuals who have impaired host defenses?

(A) Syphilis
(B) Cholera
(C) Pneumococcal pneumonia
(D) Tetanus
(E) Malaria

146. Both the Wassermann test and the VDRL test are flocculation tests for syphilis antibody. The Wassermann test originally was designed as

(A) a precipitin test
(B) a complement-fixation test
(C) a hemolysis test
(D) an agglutinin test
(E) an opsonin test

147. Granulomatosis infantiseptica, a disease characterized by meningeal necrosis, causes infant death. Which of the following agents is associated with this disease?

(A) *Streptococcus pyogenes*
(B) *Streptococcus faecalis*
(C) *Staphylococcus aureus*
(D) *Listeria monocytogenes*
(E) *Erysipelothrix rhusiopathiae*

148. *Staphylococcus aureus* causes a wide variety of infection, ranging from wound infection to pneumonia. Treatment of *S. aureus* infection with penicillin is often complicated by the

(A) inability of penicillin to penetrate the membrane of *S. aureus*
(B) production of penicillinase by *S. aureus*
(C) production of penicillin acetylase by *S. aureus*
(D) allergic reaction caused by staphylococcal protein
(E) lack of penicillin binding sites on *S. aureus*

149. A man who recently had a tooth extracted comes to a hospital's emergency room complaining of a fluctuant mass in his jaw. Gram stain of aspirated pus reveals many delicate, branching, gram-positive rods. The most likely microbiologic diagnosis is infection with a species of

(A) *Nocardia*
(B) *Actinomyces*
(C) *Mycobacterium*
(D) *Bifidobacterium*
(E) *Eubacterium*

150. Bacterial meningitis in children between the ages of six months and two years most commonly is caused by

(A) *Neisseria meningitidis*
(B) *Haemophilus influenzae*
(C) *Streptococcus pyogenes*
(D) *Streptococcus pneumoniae*
(E) *Klebsiella pneumoniae*

151. The bacterium that most commonly causes puerperal sepsis is

(A) *Treponema pallidum*
(B) *Neisseria gonorrhoeae*
(C) *Clostridium perfringens*
(D) *Streptococcus pneumoniae*
(E) *Streptococcus pyogenes*

152. Symptoms of *Clostridium botulinum* food poisoning include double vision, inability to speak, and respiratory paralysis. These symptoms are consistent with

(A) invasion of the gut epithelium by *C. botulinum*
(B) secretion of an enterotoxin
(C) endotoxin shock
(D) ingestion of a neurotoxin
(E) activation of cyclic AMP

153. Pneumococci have an antigen that determines both virulence and serotype. This antigen is a

(A) thermolabile leukocidin
(B) flagellar carbohydrate
(C) nucleoprotein derivative
(D) somatic carbohydrate
(E) capsular polysaccharide

154. Group A beta-hemolytic streptococci cause both skin infection and
pharyngitis. Which of the following
statements is associated with streptococcal skin infection?

(A) Common sequelae include rheumatic fever
(B) Common sequelae include acute
glomerulonephritis
(C) Most streptococcal skin infections
cause a rise in antistreptolysin
O (ASO) titer
(D) Clinically, infection produces
deep, suppurating ulcers
(E) The causative organism cannot
be cultured readily

155. In individuals who have sickle-
cell anemia, osteomyelitis usually is
associated with which of the following organisms?

(A) *Micrococcus*
(B) *Escherichia*
(C) *Pseudomonas*
(D) *Salmonella*
(E) *Streptococcus*

156. Oroya fever, a disease of western
South America, is caused by

(A) *Bartonella bacilliformis*
(B) *Pseudomonas pseudomallei*
(C) *Bacteroides melaninogenicus*
(D) *Pasteurella multocida*
(E) *Brucella canis*

157. The etiologic agent of tetanus is

(A) a species of *Streptococcus*
(B) a species of *Corynebacterium*
(C) a species of *Clostridium*
(D) a species of *Yersinia*
(E) ferric oxide (rust)

158. *Corynebacterium diphtheriae*,
the causative agent of diphtheria, is
best characterized by which of the
following statements?

(A) It produces an enterotoxin that
is absorbed in the gut
(B) It is isolated routinely from the
blood of infected patients
(C) It produces a toxin that blocks
protein synthesis in host cells
(D) It is resistant to most antibiotics
(E) It is part of the normal flora of
the mouth

159. Which of the following statements about "viridans" streptococci,
which include *Streptococcus mutans*,
is true?

(A) They produce beta-hemolysis
on blood agar
(B) They are not bile-soluble
(C) They are not part of the normal
flora of the upper respiratory
tract
(D) They are part of the normal flora
of the urinary tract
(E) They often settle on normal
heart valves

160. Yaws is an endemic tropical infection caused by *Treponema pertenue*. Which of the following procedures would prove most valuable in the diagnosis of yaws?

(A) A Frei test
(B) Funduscopic examination
(C) Darkfield microscopy of the lesions
(D) Blood culture in embryonated eggs
(E) Gram stain of the lesion

161. A 55-year-old man is admitted to a hospital because of a temperature of 38.9°C (102°F), chest pain, and a dry cough. A thoracentesis is performed, and culture of the fluid taken from the man's chest reveals pleomorphic gram-negative rods able to grow only on a Mueller-Hinton agar plate supplemented with the nutrient Isovitalex. This organism most likely is

(A) *Klebsiella pneumoniae*
(B) *Mycoplasma pneumoniae*
(C) *Legionella pneumophila*
(D) *Chlamydia trachomatis*
(E) *Staphylococcus aureus*

162. In infants, pneumothorax, pneumatocele, and empyema are frequent complications of pneumonia caused by

(A) *Haemophilus*
(B) *Staphylococcus*
(C) *Klebsiella*
(D) *Mycoplasma*
(E) *Streptococcus*

163. A hyperemic edema of the larynx and epiglottis rapidly leading to respiratory obstruction in young children is most likely to be caused by

(A) *Klebsiella pneumoniae*
(B) *Mycoplasma pneumoniae*
(C) *Neisseria meningitidis*
(D) *Haemophilus influenzae*
(E) *Haemophilis hemolyticus*

164. A survey of 100 seemingly healthy people revealed that 10 harbored *Neisseria meningitidis* in their nasopharynx. The most likely conclusion from this survey is that

(A) many persons have subclinical *Neisseria* pharyngitis
(B) an epidemic of neisserial meningitis is likely
(C) the laboratory was in error and the isolates were probably *N. sicca*
(D) approximately 10 percent of normal healthy persons carry *N. meningitidis*
(E) approximately 90 percent of normal healthy individuals have protective antibodies against *N. meningitidis*

165. Histologic examination of a biopsied genital lesion reveals Donovan bodies. The disease is most likely to be

(A) secondary syphilis
(B) chancroid
(C) lymphogranuloma venereum
(D) granuloma inguinale
(E) genital tuberculosis

166. Gram stain of the lesions of Vincent's stomatitis would be expected to reveal which of the following pairs of organisms?

(A) *Borrelia* and *Fusobacterium*
(B) *Fusobacterium* and *Leptospira*
(C) *Borrelia* and alpha-hemolytic *Streptococcus*
(D) *Treponema* and *Eikenella*
(E) *Escherichia* and *Eikenella*

167. *Bordetella* is divided into three species: *B. pertussis, B. parapertussis,* and *B. bronchiseptica.* Which of the following characteristics is common to all three species?

(A) Coccoid morphology
(B) Requirement for X and V factors
(C) Gram-negative staining
(D) Causative agent of whooping cough
(E) No growth on Bordet-Gengou agar

168. Acute hematogenous osteomyelitis is caused most often by

(A) *Proteus mirabilis*
(B) *Streptococcus faecalis*
(C) *Staphylococcus epidermidis*
(D) *Staphylococcus aureus*
(E) *Escherichia coli*

169. Diphtheria toxin is produced only by those strains of *Corynebacterium diphtheriae* that are

(A) glucose fermenters
(B) lysogenic for β-prophage
(C) sucrose fermenters
(D) of the mitis strain
(E) encapsulated

170. In cases of suspected shigellosis, the best method of obtaining a specimen for culture of the causative organism is by

(A) rectal swab
(B) stool culture
(C) urine culture
(D) venipuncture
(E) spinal tap

171. The most common site of asymptomatic gonococcal infection in women is the

(A) vagina
(B) myometrium
(C) urethra
(D) fallopian tubes
(E) endocervix

172. Primary syphilis is characterized by

(A) initial lesions appearing three months after infection
(B) a positive darkfield examination
(C) a negative VDRL test
(D) gummas
(E) tabes dorsalis

173. Secondary syphilis is characterized by

(A) penile chancres
(B) a rash only on the palms of the hands
(C) a positive serum VDRL test
(D) a positive cerebrospinal fluid VDRL test
(E) a negative fluorescent treponemal antibody-absorption test

174. Food poisoning that produces
gastrointestinal symptoms approximately one to two hours after eating
is most likely to be due to

(A) *Salmonella enteritidis*
(B) *Streptococcus faecalis*
(C) *Clostridium perfringens*
(D) *Staphylococcus aureus*
(E) *Shigella sonnei*

DIRECTIONS: Each question below contains four suggested answers of which **one** or **more** is correct. Choose the answer:

A	if	**1, 2, and 3**	are correct
B	if	**1 and 3**	are correct
C	if	**2 and 4**	are correct
D	if	**4**	is correct
E	if	**1, 2, 3, and 4**	are correct

175. The principal pathogens of bacterial pneumonia include

(1) *Staphylococcus*
(2) *Enterobacter*
(3) *Streptococcus*
(4) *Mycobacterium*

176. In addition to gonorrhea, which of the following diseases are considered to be transmitted venereally?

(1) Condyloma lata
(2) Granuloma inguinale
(3) Herpes progenitalis
(4) Chlamydial lymphogranuloma

177. Leprosy is a worldwide public-health problem. *Mycobacterium leprae*, the causative organism, is

(1) acid-fast
(2) grown easily on Sabouraud agar
(3) capable of causing a false positive VDRL test
(4) demonstrable by serologic tests

178. Which of the following diseases are caused by spirochetes?

(1) Yaws
(2) Pinta
(3) Infectious jaundice
(4) Relapsing fever

179. Recent infection with group A beta-hemolytic streptococci may produce an increase in antibody titers to

(1) erythrogenic toxin
(2) DNase B
(3) hyaluronidase
(4) streptolysin S

180. Which of the following statements about deoxycholate agar, which is used for culturing enterobacteria, are true?

(1) It contains neutral red as an indicator
(2) It contains glucose
(3) It inhibits the growth of gram-positive organisms
(4) Lactose fermenters grow as colorless colonies on it

181. The usual laboratory criteria for assessing the pathogenicity of *Staphylococcus aureus* include

(1) lactose fermentation
(2) hemolysis
(3) red pigment production
(4) coagulase activity

182. A mutation that causes a strain of *Neisseria gonorrhoeae* to lose its pili would result in

(1) inability to utilize glucose
(2) failure to grow on Thayer-Martin agar
(3) loss of Gram-stain reactivity
(4) altered virulence

183. Current detection and isolation methods for *Legionella pneumophila* in pleural fluid include

(1) direct fluorescent staining
(2) culture of the organism
(3) Giemsa staining
(4) Gram staining

184. Intermittent fever is a common clinical sign. Which of the following organisms can produce intermittent fever in patients?

(1) *Borrelia recurrentis*
(2) *Brucella melitensis*
(3) *Mycobacterium tuberculosis*
(4) *Yersinia pestis*

185. Characteristics of *Salmonella-Shigella* (SS) agar include

(1) potentiation of the growth of coliform colonies
(2) suppression of the growth of gram-positive organisms
(3) production of black colonies of *Salmonella*
(4) the presence of bile salts

186. The growth of *Neisseria* in culture is enhanced by

(1) incubation at 50°C
(2) increased carbon dioxide tension
(3) the addition of fatty acids to the medium
(4) the inclusion of blood proteins in the medium

187. The most common disease caused by group A beta-hemolytic streptococci is pharyngitis. These streptococci also can cause

(1) erysipelas
(2) impetigo
(3) glomerulonephritis
(4) puerperal fever

188. The normal microbial flora of the skin of healthy individuals includes

(1) aerobic diphtheroid bacilli
(2) anaerobic diphtheroid bacilli
(3) nonhemolytic staphylococci
(4) Enterobacteriaceae

189. Experimental alteration of the vaginal pH from 7.4 to 5.6 would result in the appearance of which of the following organisms?

(1) Group B streptococci
(2) Clostridia
(3) Coliforms
(4) Lactobacilli

SUMMARY OF DIRECTIONS

A	B	C	D	E
1, 2, 3 only	1, 3 only	2, 4 only	4 only	All are correct

190. *Francisella tularensis*, the agent of tularemia, is characterized by

(1) a requirement for cysteine
(2) an inability to form spores
(3) pleomorphism
(4) person-to-person transmission

191. Although more than 2,000 serotypes of *Salmonella* have been recognized, there are relatively few biotypes (species). These species include

(1) *S. enteritidis*
(2) *S. typhi*
(3) *S. choleraesuis*
(4) *S. derby*

192. Species of the genus *Bacillus* that are highly pathogenic for humans include

(1) *B. cereus*
(2) *B. subtilis*
(3) *B. megaterium*
(4) *B. anthracis*

193. Protein toxin production is characteristic of

(1) *Corynebacterium diphtheriae*
(2) *Clostridium tetani*
(3) *Staphylococcus aureus*
(4) *Mycobacterium tuberculosis*

194. Motility is an important distinguishing feature among microorganisms. Which of the following bacteria are **nonmotile**?

(1) Streptococci
(2) Staphylococci
(3) Klebsiellae
(4) Clostridia

195. A gram-negative coccus resembling *Neisseria* is isolated from cerebrospinal fluid. Initial serologic testing is negative with *N. meningitidis* antiserum groups A, B, and C. Subsequent identification steps should include

(1) testing with X, Y, and Z antisera
(2) an oxidase test
(3) carbohydrate utilization tests
(4) growth in an anaerobic jar

196. Endotoxin produced by gram-negative bacteria can cause

(1) hemorrhagic tissue necrosis
(2) disseminated intravascular coagulation (DIC)
(3) the Shwartzman phenomenon
(4) fever

197. Gram-negative rods are the etiologic agents in which of the following diseases?

(1) Neisserial meningitis
(2) Ludwig's angina
(3) Ophthalmia neonatorum
(4) Chancroid

198. Relapsing fever, caused by *Borrelia recurrentis*, is characterized by

(1) an incubation period of three to ten days
(2) recurring bouts of progressively higher temperature
(3) intense headaches
(4) diarrhea

199. Which of the following tests would help to distinguish *Streptococcus pneumoniae* from other alpha-hemolytic streptococci?

(1) Bile solubility
(2) Optochin sensitivity
(3) Mouse virulence
(4) Catalase activity

200. Bacterial capsules have an anti-phagocytic function. Which of the following organisms do NOT have a capsule?

(1) *Streptococcus pneumoniae*
(2) *Klebsiella pneumoniae*
(3) *Bacillus anthracis*
(4) *Corynebacterium diphtheriae*

201. An isolate from an anaerobic blood culture bottle is a gram-negative rod. It is resistant to both gentamicin and penicillin. Confirmation of this isolate as *Bacteroides fragilis* can be accomplished by

(1) gas chromatographic analysis of metabolic by-products
(2) proof of its anaerobic nature
(3) a study of its biochemical reactivity
(4) a test for clindamycin resistance

202. Disorders caused by staphylococci include

(1) carbuncles
(2) scarlet fever
(3) meningitis
(4) pertussis

203. Vaccines against plague and tularemia can be prepared from

(1) avirulent live bacteria
(2) heat-killed suspensions of virulent bacteria
(3) formalin-inactivated suspensions of virulent bacteria
(4) chemical fractions of the causative bacilli

204. A 48-year-old man known to be an alcoholic is admitted to a hospital with a diagnosis of pneumonia. Gram stain of his sputum reveals encapsulated gram-negative rods and many polymorphonuclear leukocytes. Which of the following biochemical characteristics and antibiotic-susceptibility test results would be likely for this isolate?

(1) Indole-negative
(2) Carbenicillin-resistant
(3) Ampicillin-resistant
(4) Citrate-negative

205. A newborn infant has signs and symptoms of *Listeria* meningitis. Which of the following bacteriologic characteristics are suggestive of *Listeria monocytogenes*?

(1) Beta-hemolytic on blood agar
(2) Motile (in a tumbling fashion)
(3) Gram-positive
(4) Catalase-positive

SUMMARY OF DIRECTIONS				
A	B	C	D	E
1, 2, 3 only	1, 3 only	2, 4 only	4 only	All are correct

206. The ability to distinguish among the species of the genus *Shigella* may have important clinical advantages. Testing for which of the following biochemical characteristics would be useful in identifying a particular species of *Shigella*?

(1) Ortho-nitrophenyl-beta-D-galac-topyranoside (ONPG) hydrolysis
(2) Presence of ornithine decarboxy-lase
(3) Xylose fermentation
(4) Mannitol fermentation

207. Which of the following would distinguish between *Nocardia* and *Actinomyces*?

(1) Pathogenicity
(2) Gram-staining characteristics
(3) Lack of growth in a carbon-dioxide-enriched environment
(4) Lack of growth in an aerobic environment

DIRECTIONS: The groups of questions below consist of lettered choices followed by several numbered items. For each numbered item select the **one** lettered choice with which it is **most** closely associated. Each lettered choice may be used once, more than once, or not at all.

Questions 208-212

For each antimicrobial drug below, choose its principal side-effect.

(A) Gastrointestinal upset
(B) Peripheral neuritis
(C) Neurotoxicity
(D) Optic neuritis
(E) Eighth-cranial-nerve toxicity

208. Cycloserine

209. Ethambutol

210. Isoniazid (INH)

211. Para-aminosalicylic acid (PAS)

212. Streptomycin

Questions 213-217

For each bacterium listed below, choose its preferred isolation medium.

(A) Sheep blood agar
(B) Löffler's medium
(C) Thayer-Martin agar
(D) Thiosulfate-citrate bile salts medium
(E) Löwenstein-Jensen agar

213. *Neisseria gonorrhoeae*

214. *Vibrio cholerae*

215. *Mycobacterium tuberculosis*

216. *Corynebacterium diphtheriae*

217. *Staphylococcus aureus*

Questions 218-221

For each organism below, choose the description with which it is most likely to be associated.

(A) This small, strongly urease-positive gram-negative rod is a common inhabitant of the canine respiratory tract and an occasional pathogen for humans

(B) This small gram-negative rod pits agar, grows both in carbon dioxide and under anaerobic conditions, and is part of the normal oral-cavity flora

(C) This small gram-negative rod typically infects cattle, requires 5 to 10% carbon dioxide for growth, and is inhibited by the dye thionine

(D) This small gram-negative rod typically is found in infected animal bites in humans and can cause hemorrhagic septicemia in animals

(E) This small gram-negative rod manifests different biochemical and physiologic characteristics, depending on growth temperature, and causes a spectrum of human disease, most commonly mesenteric lymphadenitis

218. *Yersinia enterocolitica*

219. *Brucella abortus*

220. *Bordetella bronchiseptica*

221. *Pasteurella multocida*

Questions 222-225

Match the following.

(A) A facultative anaerobe that often inhabits the buccal mucosa early in a neonate's life and can cause bacterial endocarditis

(B) A beta-hemolytic organism that cause a diffuse, rapidly spreading cellulitis

(C) An anaerobic, filamentous bacterium that often causes cervicofacial osteomyelitis

(D) A facultative anaerobe that is highly cariogenic and sticks to teeth by synthesis of a dextran

(E) A facultatively anaerobic rod-shaped bacterium that sticks to teeth and is cariogenic

222. *Streptococcus mutans*

223. *Streptococcus salivarius*

224. *Actinomyces viscosus*

225. *Actinomyces israelii*

Bacteriology

Answers

115. The answer is E. *(Jawetz, ed 13. pp 9-19.)* Bacterial cell envelopes consist of both the cell wall and cell membrane. The cell envelope of gram-negative bacteria contains lipoprotein, lipopolysaccharide, and peptidoglycan molecules; the polysaccharide component of the lipopolysaccharide is the O antigen, the chief surface antigen of these bacteria. Gram-positive organisms, on the other hand, contain large amounts of teichoic acids, which are important surface antigens for these bacteria. The cell wall of gram-positive bacteria acts as a barrier to the extraction of crystal violet-iodine complex by alcohol; this property is the basis of the Gram stain. Large charged molecules do not freely diffuse through the cytoplasmic membranes of either gram-positive or gram-negative cells.

116. The answer is C. *(Davis, ed 2. p 523.)* In-the presence of specific antibody and complement, red blood cells and most gram-negative organisms undergo lysis. Thus, the gram-negative *Klebsiella pneumoniae* would be destroyed by complement-antibody complexes. However, for poorly understood reasons, neither gram-positive organisms (e.g., streptococci, clostridia, corynebacteria) nor mycobacteria are lysed under these conditions.

117. The answer is B. *(Lennette, ed 2. p 908.)* Mannitol salt agar is a selective medium for staphylococci. Growth of *Escherichia coli* on mannitol salt agar is usually due to a low salt content of the medium. Unlike other bacteria, staphylococci grow readily in 7.5% sodium chloride.

118. The answer is A. *(Jawetz, ed 13. pp 234-235.)* Serologic tests for syphilis, which is caused by the spirochete *Treponema pallidum*, fall into two groups. The VDRL, Wassermann, and automated reagin tests detect the presence of reagin, an antibody-like compound produced during spirochetal infection. Tests for the presence of specific antitreponemal antibodies include the fluorescent treponemal antibody-absorption (FTA-ABS) test. The Frei test is a skin test used in the diagnosis of lymphogranuloma venereum, a chlamydial disease.

119. The answer is B. *(Jawetz, ed 13. pp 180-181.)* The Dick test can be used to show an individual's susceptibility to the beta-hemolytic streptococcal erythrogenic toxin that causes the rash of scarlet fever. Tuberculosis is associated with the purified protein derivative (PPD-S) test, lymphogranuloma venereum with the Frei test, and sarcoidosis with the Kveim test. The Dick test is no longer widely used.

120. The answer is A. *(Jawetz, ed 13. pp 181-182.)* Clinically, the child described in the question has group B streptococcal meningitis. Although most streptococcal disease in humans is caused by group A streptococci, newborn infants can be infected with group B streptococci, which normally reside in the vagina. After Gram stain and culture have been performed, identification should include either a test for hippurate hydrolysis or the "CAMP" test and confirmation of the serogroup of the organism. The CAMP test determines the presence of a substance that is secreted by group B streptococci and that enhances hemolysis production by *Staphylococcus aureus*; other streptococci rarely can elicit a positive CAMP test.

121. The answer is C. *(Youmans, p 757.)* The interpretation of quantitative antimicrobial susceptibility tests is based on both the minimum inhibitory concentration (MIC) and the achievable blood level of a given antibiotic. An MIC greater than the achievable blood concentration of an antibiotic suggests resistance. An MIC at or near the achievable level is equivocal, and an MIC significantly lower than the achievable level—say, by 75 percent—suggests susceptibility of the isolate to the antibiotic being tested.

122. The answer is C. *(Davis, ed 2. pp 45-46.)* *Enterobacter*, certain other Enterobacteriaceae, and *Bacillus* ferment pyruvate to form butylene glycol. On exposure to air, this compound is oxidized to acetoin. The presence of acetoin can be determined readily by the Voges-Proskauer test, useful in differentiating between *Escherichia coli* and *Enterobacter*.

123. The answer is C. *(Volk, p 318.)* *Pasteurella multocida* is part of the normal mouth flora of dogs and cats. Consequently, many animal bites become infected with this microorganism. *P. multocida* is susceptible to penicillin.

124. The answer is B. *(Thorn, ed 8. pp 912-915.)* Dapsone (diaminodiphenylsulfone) is considered at present to be the drug of choice for treating individuals who have leprosy. A daily adult maintenance dose of 50 mg, administered over several months, should eradicate the basic disease. However, leprosy bacilli can persist in a host for many years; as a consequence, dapsone should be given for six to ten years after the disappearance of the bacilli from skin smears. Recent information has shown that rifampin may be the future drug of choice for leprosy.

125. The answer is C. *(Youmans, p 466.)* The onset of clinical gonococcal infection is characterized by fever, chills, and polyarthralgia. These symptoms may be followed by acute arthritis of one or more joints, acute salpingitis, and sterility. Gram-negative diplococci occasionally can be seen in a Gram stain of synovial fluid.

126. The answer is B. *(Burrows, ed 20. p 643.)* Löffler's medium is used to culture diphtheria bacilli (*Corynebacterium diphtheriae*). It is especially useful in throat cultures, because the medium does not support the growth of other common pharyngeal organisms (pneumococci and streptococci). Corynebacteria can grow on the majority of ordinary culture media.

127. The answer is B. *(Jawetz, ed 13. p 185.)* When antiserum to capsular polysaccharide is added to a slide preparation of organisms, the capsule appears swollen microscopically. This phenomenon, the quellung reaction, can be used for rapid identification of certain pneumococci. Whether the capsule actually "swells" is questionable; however, visualization of the capsule certainly is improved following the specific antigen-antibody reaction.

128. The answer is C. *(Jawetz, ed 13. p 200.)* In unimmunized individuals diphtheria toxin (but **not** toxoid) produces an inflammatory reaction over the course of several days. Circulating antibodies prevent this reaction in persons immune to diphtheria. A few individuals are hypersensitive to extraneous factors in the toxin and toxoid and develop an inflammatory pseudoreaction that appears in both arms simultaneously and disappears within three days. An initial reaction in both arms that subsides quickly only at the toxoid site demonstrates hypersensitivity and susceptibility.

129. The answer is C. *(Jawetz, ed 13. p 258.)* *Acinetobacter*, a gram-negative aerobe, is an opportunistic pathogen. Infection usually develops in immunologically compromised hosts. *Acinetobacter, Pseudomonas, Moraxella,* and *Flavobacterium* are water-borne organisms often isolated from hospital appliances, including mist tents and humidifiers.

130. The answer is B. *(Jawetz, ed 13. p 181.)* The Schultz-Carlton reaction can be used to show that an individual's rash is due to scarlet fever. In this test, antibody to erythrogenic toxin is injected directly into the rash. If the rash fades or disappears (i.e., if the toxin has been neutralized), then the diagnosis of scarlet fever can be made with certainty.

131. The answer is C. *(Jawetz, ed 13. pp 234-235.)* In men, the appearance of a hard chancre on the penis characteristically indicates syphilis. Even though the chancre does not appear until the infection is two or more weeks old, the VDRL test for syphilis still can be negative despite the presence of a chancre (the VDRL test may not become positive for two or three weeks after initial infection). However, a lesion suspected of being a primary syphilitic ulcer should be examined by darkfield microscopy, which can reveal motile treponemes.

132. The answer is A. *(Volk, pp 262-263.)* The Lancefield classification of streptococci is based on the presence of a C-carbohydrate in the cell wall. Typing within a group, however, is mediated by other substances. In group A streptococci, the type-specific antigen is a protein (M-protein); in other streptococci, the type-specific substance may be a carbohydrate.

133. The answer is E. *(Burrows, ed 20. pp 664, 727, 880, 882.)* Granulomatous lesions are circumscribed nodular lesions characterized by the presence of macrophages. They may persist for a long time as sites of smoldering inflammation. Granulomas develop in a variety of diseases, including tuberculosis, sarcoidosis, coccidioidomycosis, and cat-scratch fever. Chemical and mineral irritation also can produce granulomatous reactions.

134. The answer is E. *(Davis, ed 2. pp 635-637. Jawetz, ed 13. p 210.)* Exotoxins are diffusible substances elaborated chiefly by gram-positive organisms, whereas endotoxins are cell-wall components of certain gram-negative bacteria. The exotoxins of *Clostridium tetani* and *C. botulinum* act directly on the nervous system. The "shiga toxin" of *Shigella* dysentery acts on the smaller cerebral blood vessels. Diphtheria exotoxin affects body cells in general. Disseminated intravascular coagulation (DIC) results from many conditions, including the action of gram-negative bacterial endotoxin on the intrinsic clotting system.

135. The answer is A. *(Jawetz, ed 13. p 223.)* Brucellae are small, aerobic, gram-negative coccobacilli. Of the four well-characterized species of *Brucella*, only one—*B. melitensis*—characteristically infects both goats and humans. Brucellosis may be associated with gastrointestinal and neurologic symptoms, lymphadenopathy, splenomegaly, hepatitis, and osteomyelitis.

136. The answer is D. *(Jawetz, ed 13. p 187.)* Pathogenic neisseriae (*Neisseria meningitidis, N. gonorrhoeae*) will not grow on plain agar; they grow best on blood-enriched plates in the presence of 10% carbon dioxide. *Branhamella catarrhalis, N. flavescens,* and *N. sicca* all grow on plain nutrient agar. *N. meningitidis* (strain **A** in the question) produces acid from maltose and dextrose, while *N. gonorrhoeae* (strain **B**) ferments only dextrose. Strain **C** could be either *B. catarrhalis* or *N. flavescens. N. sicca* produces acid from sucrose, maltose, and dextrose.

137. The answer is E. *(Volk, p 301.)* Cholera is a toxicosis. The mode of action of cholera toxin is to stimulate the activity of adenyl cyclase, an enzyme that converts ATP to cyclic AMP. Cyclic AMP stimulates the secretion of chloride ion, and affected individuals lose copious amounts of fluid. A drug that inhibits adenyl cyclase thus might block the effect of cholera toxin.

138. The answer is E. *(Davis, ed 2. p 725.)* Subacute bacterial endocarditis is an insidious infection that occurs on heart valves previously damaged by disease, such as rheumatic fever. The most common causative agent is α-hemolytic *Streptococcus* (viridans streptococci), a normal oral-cavity inhabitant that can infect the heart after minor trauma, dental manipulation, or other procedures. In contrast, acute bacterial endocarditis is a rapidly progressive infection that occurs on undamaged valves and may be caused by staphylococci, enterococci, and pneumococci.

139. The answer is A. *(Davis, ed 2. pp 758-759, 776-777.)* Both shigellae and salmonellae are gram-negative enteric bacilli that ferment mannitol. Usually, both are methyl-red-positive and urease-negative and neither produces a Voges-Proskauer reaction. Shigellae are distinguished from salmonellae by their lack of motility.

140. The answer is B. *(Jawetz, ed 13. p 186.)* The clinical syndrome described in the question is classic for pneumococcal pneumonia, which is caused by *Streptococcus pneumoniae*. Frequently, an upper respiratory tract infection precedes the development of pneumonia in hosts, such as alcoholics, whose natural resistance to pneumococcal infection is lowered. With proper antimicrobial therapy (the treatment of choice is penicillin), patients may become afebrile within 48 hours and further consolidation of the lungs can be avoided.

141. The answer is A. *(Davis, ed 2. p 784.)* *Pseudomonas aeruginosa* may cause severe and potentially fatal infections in patients with extensive burns. It also may be responsible for urinary tract infections, septicemia, abscesses, and meningitis. *P. aeruginosa* flourishes particularly well in patients being treated with corticosteroids or antibiotics. The use of a Wood's ultraviolet lamp usually can differentiate this organism from other gram-negative bacteria. Most pseudomonads produce the characteristic blue-green pigment pyocyanin.

142. The answer is E. *(Thorn, ed 8. pp 830-831.)* Although part of the normal intestinal flora, strains of *Escherichia coli* become pathogenic when they invade other organ systems. More than 50 percent of *E. coli* infections begin in the urinary tract, and 75 percent of urinary tract infections are caused by *E. coli*. These organisms, however, can produce infection in any area of the body.

143. The answer is A. *(Jawetz, ed 13. p 176.)* Certain strains of staphylococci elaborate an enterotoxin that is frequently responsible for food poisoning. Typically, the toxin is produced when staphylococci grow on foods rich in carbohydrates and is present in the food when it is consumed. The resulting gastroenteritis is dependent only on the ingestion of toxin and not upon bacterial multiplication in the gastrointestinal tract.

144. The answer is C. *(Jawetz, ed 13. p 198.)* Corynebacteria may be recognized microscopically by the presence of Babès-Ernst bodies, which are metachromatic granules made visible by aniline staining. These granules give the rods a beaded appearance. Babès-Ernst bodies are granules of polymerized polyphosphates.

145. The answer is C. *(Harvey, ed 19. p 1231.)* When an individual suffers repeated attacks of the same type of pneumococcal pneumonia, dysgammaglobulinemia, especially multiple myeloma, should be suspected. Recurrent infection is probably due to the inability of the host to produce anticapsular antibody, which increases the phagocytosis of the bacteria in the lungs. Impaired pulmonary function and alcoholic intoxication also tend to increase the risk of pneumococcal pneumonia due to a delay in the appearance of polymorphonuclear leukocytes.

146. The answer is B. *(Davis, ed 2. pp 885-886.)* The original Wassermann test was a complement-fixation test. However, it was found that false positive reactions occurred in association with a variety of other conditions, including malaria, leprosy, lupus erythematosus, and polyarteritis nodosa. Variations of the Wassermann test based on flocculation reactions have been developed in the hope that a more specific test will emerge.

147. The answer is D. *(Davis, ed 2. pp 946-948.)* *Listeria monocytogenes,* a short, gram-positive, nonspore-forming rod, is the cause of granulomatosis infantiseptica (perinatal listeric septicemia), which is associated with an often asymptomatic, low-grade uterine infection in pregnant women. In affected neonates, however, mortality is high; focal necrosis of the meninges, liver, and other organs is characteristic. Intrauterine infection with *Listeria* also may result in stillbirth or abortion.

148. The answer is B. *(Jawetz, ed 13. pp 175-179.)* Staphylococci are grampositive, nonspore-forming cocci. Clinically, their antibiotic resistance poses major problems. Many strains produce β-lactamase (penicillinase), an enzyme that destroys penicillin by opening the lactam ring. Drug resistance, mediated by plasmids, may be transferred by transduction.

149. The answer is B. *(Volk, pp 388-389.)* *Actinomyces israelii* is a grampositive, branched, fungus-like bacterium. It characteristically produces face and neck abscesses, which tend to drain spontaneously and form chronic sinus tracts. Diagnosis may be made by the finding of typical sulfur granules in the pus or by anaerobic culture on blood agar plates and supplemented thioglycollate medium.

150. The answer is B. *(Davis, ed 2. p 795.)* Except during a meningococcal epidemic, *Haemophilus influenzae* is the most common cause of bacterial meningitis in children. Most cases are secondary to respiratory tract infections or otitis media. *H. influenzae, Neisseria meningitidis,* and *Streptococcus pneumoniae* account for 80 to 90 percent of all cases of bacterial meningitis.

151. The answer is E. *(Jawetz, ed 13. p 182.)* Puerperal fever is a uterine infection that can follow childbirth or abortion. It commonly is caused by streptococci (especially *Streptococcus pyogenes*), staphylococci, or *Escherichia coli*. Rarely, clostridia may be the causative agent, especially following criminal abortions.

152. The answer is D. *(Jawetz, ed 13. pp 193-195.)* *Clostridium botulinum* growing in food produces a potent neurotoxin that when ingested by humans produces diplopia, dysphagia, respiratory paralysis, and speech difficulties. The toxin is thought to act by blocking the action of acetylcholine at neuromuscular junctions. Botulism is associated with a high rate of mortality; fortunately, *C. botulinum* infection in humans is rare.

153. The answer is E. *(Davis, ed 2. pp 697-698. Jawetz, ed 13. pp 178-180.)* The virulence of pneumococci is dependent upon the presence of a capsule. Capsular polysaccharides inhibit phagocytosis and destruction of the organism. In serum that contains specific anticapsular antibodies, this protective function is lost. Pneumococci may be serotyped on the basis of their capsular polysaccharides.

154. The answer is B. *(Youmans, pp 212-232.)* Skin streptococci are usually nephritogenic and not rheumatogenic. Not uncommonly, patients who have acute glomerulonephritis do not show an elevated titer of antistreptolysin O (ASO) but do have antibody titers to other streptococcal antigens, such as diphosphopyridine nucleotidase and deoxyribonuclease (DNase). Streptococcal skin infection (impetigo) most often affects young children, is highly contagious, and produces superficial blisters.

155. The answer is D. *(Hoeprich, ed 2. p 1133.)* Many types of infection, notably respiratory tract infections and osteomyelitis, are common in individuals who have sickle-cell anemia. For unknown reasons, *Salmonella* is implicated frequently in these infections. Osteomyelitis in other individuals is caused most often by *Staphylococcus*.

156. The answer is A. *(Jawetz, ed 13. p 258.)* Oroya fever, a disease found only in the Andes Mountains of Peru, Colombia, and Ecuador, is caused by *Bartonella bacilliformis*. The infection is characterized by severe anemia, hepatosplenomegaly, and hemorrhage. The same organism also can cause a benign skin infection called verruga peruana.

157. The answer is C. *(Jawetz, ed 13. pp 194-195.)* *Clostridium tetani*, an anaerobic, gram-positive rod, is the causative organism of tetanus. It is isolated often from soil and animal feces and produces a localized infection in contaminated wounds, burns, and surgical incisions. *C. tetani* elaborates a powerful neurotoxin that causes violent tonic contractions of voluntary muscles. The common name for tetanus, "lockjaw," derives from the intense spasms produced in the masseter muscles.

158. The answer is C. *(Jawetz, ed 13. pp 198-201.)* Diphtheria, a disease caused by *Corynebacterium diphtheriae*, usually begins as a pharyngitis associated with "pseudomembrane" formation and lymphadenopathy. Growing organisms produce a potent exotoxin that is absorbed in mucous membranes and causes remote damage to the liver, kidneys, and heart; the polypeptide toxin inhibits host-cell protein synthesis. Although *C. diphtheriae* may infect the skin, it rarely invades the bloodstream and never actively invades deep tissue.

159. The answer is B. *(Jawetz, ed 13. pp 181-182.)* "Viridans" streptococci produce alpha-hemolysis, but not beta-hemolysis, on blood agar plates. They are normal inhabitants of the oral cavity and frequently become disseminated during dental manipulation. The resulting bacteremia may seed abnormal heart valves and produce endocarditis. Unlike pneumococci, they are not soluble in bile. The term "viridans" designates a large group of streptococci, comprising at least 10 species.

160. The answer is C. *(Jawetz, ed 13. p 236.)* Yaws is an endemic tropical infection caused by *Treponema pertenue*. The organism is closely related to the spirochete that causes syphilis and will give a biologic true-positive VDRL test. It can be observed in the lesions by darkfield microscopy. Unlike syphilis, yaws is not transmitted venereally but by person-to-person contact, usually among children.   *and fly bites*

161. The answer is C. *(Fraser, N Engl J Med 297 [1977]:1189-1197. McDade, N Engl J Med 297 [1977]:1197-1203.)* The symptoms of Legionnaires' disease are similar to those of *Mycoplasma pneumoniae* pneumonia and influenza. Affected individuals are moderately febrile, complain of pleuritic chest pain, and have a dry cough. Unlike *Klebsiella* and *Staphylococcus*, *Legionella pneumophila* exhibits fastidious growth requirements, growing only on Mueller-Hinton agar supplemented with Isovitalex.

162. The answer is B. *(Thorn, ed 8. pp 805, 811.)* During the clinical course of primary staphylococcal pneumonia in infants, abscess formation and necrosis can occur throughout the lung parenchyma. These abscesses can rupture into bronchial walls or the pleural cavity, producing pyopneumothorax or pneumatoceles. Surgical intervention often is required.

163. The answer is D. *(Jawetz, ed 13. pp 229-230.)* *Haemophilus influenzae* is a gram-negative bacillus. In young children it can cause pneumonitis, sinusitis, otitis, and meningitis. Occasionally it produces a fulminative laryngotracheitis with such severe swelling of the epiglottis that tracheostomy becomes necessary. Clinical infections with this organism after the age of three years are less frequent.

164. The answer is D. *(Volk, p 279.)* Approximately 10 percent of healthy adults are carriers of *Neisseria meningitidis*. This percentage increases when people are housed in close quarters, such as military barracks. Carriage of the organism usually predisposes to the formation of protective antibody.

165. The answer is D. *(Jawetz, ed 13. p 290.)* Donovan bodies are intracellular gram-negative coccobacillary organisms thought to be the cause of granuloma inguinale. *Calymmatobacterium granulomatis*, the causative agent, is transmitted by sexual intercourse. Granuloma inguinale causes a painless, nonhealing ulceration of the genitals; the disease process may be mistaken for carcinoma.

166. The answer is A. *(Jawetz, ed 13. p 239.)* Vincent's stomatitis (trench mouth, ulcerative gingivostomatitis) is an ulcerative condition affecting the oral mucous membranes. Typically, the spirochete *Borrelia vincentii* and the bacillus *Fusobacterium plauti-vincenti (Bacteroides fusiformis)* are found in the lesion. Other bacteria, including cocci and bacilli, also may play a role.

167. The answer is C. *(Jawetz, ed 13. pp 230-232.)* *Bordetella pertussis* is the major cause of whooping cough, although *B. parapertussis* may cause a similar disease. *B. bronchiseptica* has been implicated in cases of pneumonitis. All three species, which are biochemically distinct, are small, gram-negative rods.

168. The answer is D. *(Thorn, ed 8. pp 810-811.)* *Staphylococcus aureus* is implicated in the majority of cases of acute osteomyelitis, which affects children most often. A superficial staphylococcal lesion frequently precedes the development of bone infection. In the preantibiotic era, *Streptococcus pneumoniae* was a common cause of acute osteomyelitis. *Mycobacterium tuberculosis* and gram-negative organisms are implicated less frequently.

169. The answer is B. *(Davis, ed 2. pp 684-685.)* All toxigenic strains of *Corynebacterium diphtheriae* are lysogenic for β-phage carrying the *tox* gene, which is inferred to code for the toxin molecule. The expression of this gene is controlled by the metabolism of the host bacteria. The greatest amount of toxin is produced by bacteria grown on media containing very low amounts of iron.

170. The answer is A. *(Davis, ed 2. p 778.)* Shigellae are responsible for bacillary dysentery. Although the organisms are excreted heavily in stool, they are fragile and remain viable for only a short period of time. Ideally, specimens should be obtained by rectal swabbing under direct vision through a sigmoidoscope.

171. The answer is E. *(Youmans, p 472.)* Asymptomatic gonococcal infection in women most frequently involves the endocervix. In decreasing order of occurrence, the urethra, anal canal, and pharynx also can be infected. Extension of the disease to the fallopian tubes usually is accompanied by the signs and symptoms of acute salpingitis.

172. The answer is B. *(Jawetz, ed 13. p 234.)* The typical lesion of primary syphilis is the chancre, which commonly develops within three weeks of exposure. Chancres may be found on the external genitalia, cervix, anus, or mouth. A diagnosis can be made by darkfield microscopic examination or by a positive VDRL test. Tabes dorsalis and gummas are characteristic of late syphilis.

173. The answer is C. *(Jawetz, ed 13. pp 234-235.)* Symptoms of secondary syphilis appears six to eight weeks after the chancre of primary syphilis. Symptoms include generalized rash, lymphadenopathy, iritis, arthritis, fever, and malaise. In secondary syphilis both the serum VDRL and fluorescent treponemal antibody-absorption (FTA-ABS) tests are positive. A positive cerebrospinal fluid VDRL test is a manifestation of neurosyphilis, a late stage of the disease.

174. The answer is D. *(Davis, ed 2. pp 736, 774, 839.)* Staphylococcal food poisoning is characterized by an incubation period ranging from one to eight hours. Symptoms include violent nausea, vomiting (often projectile), and diarrhea. Patients are afebrile and convalescence usually is rapid. Symptoms are due to the ingestion of a preformed enterotoxin and not to infection. Food poisoning due to salmonellae does not appear until one to three days after ingestion of the organisms; fever commonly is noted. *Clostridium perfringens* produces a gastroenteritis about 12 hours after ingestion of contaminated food.

175. The answer is B (1, 3). *(Jawetz, ed 13. pp 177, 186.)* Pneumonia may be caused by a wide variety of bacteria, fungi, and viruses. Among the more frequent bacterial agents of pneumonia are pneumococci (*Streptococcus pneumoniae*) and staphylococci. Less common causes of bacterial pneumonia are *Klebsiella*, *Enterobacter*, and *Mycobacterium*.

176. The answer is E (all). *(Jawetz, ed 13. pp 234, 250, 289-290, 447.)* The four diseases mentioned in the question are all venereal diseases. Lymphogranuloma venereum is caused by *Chlamydia*, and genital herpes (herpes progenitalis) by a herpesvirus type 2. Condyloma lata are highly infectious lesions of secondary syphilis. Granuloma inguinale is an ulcerative disease caused by the gram-negative bacterium *Calymmatobacterium (Donovania) granulomatis*.

177. The answer is B (1, 3). *(Jawetz, ed 13. pp 207-208.)* *Mycobacterium leprae* is an acid-fast bacillus. Although these mycobacteria were the first bacteria to be associated with human disease, they have never been grown on artificial media. There are no diagnostic serologic tests for *M. leprae*, although it can cause false positive VDRL tests.

178. The answer is E (all). *(Jawetz, ed 13. pp 233-239.)* Organisms of the order Spirochaetales are responsible for a variety of human diseases. Diseases caused by treponemes include yaws (*Treponema pertenue* infection), pinta *(T. carateum)*, and, of course, syphilis (*T. pallidum*). Relapsing fever is caused by *Borrelia recurrentis* and rat-bite fever by *Spirillum minus*. Diseases caused by the spirochete *Leptospira* include Weil's disease and infectious jaundice.

179. The answer is A (1, 2, 3). *(Jawetz, ed 13. pp 180-181. Youmans, p 199.)* Streptococcal infection usually is accompanied by an elevated titer of antibody to some of the enzymes produced by the organism. Among the antigenic substances elaborated by group A beta-hemolytic streptococci include erythrogenic toxin, streptodornase (streptococcal DNase), streptolysin O (a hemolysin), and hyaluronidase. Streptolysin S is a nonantigenic hemolysin.

180. The answer is B (1, 3). *(Davis, ed 2. p 759.)* Deoxycholate agar is used for the isolation of enteric bacilli from stool specimens. The medium contains bile salts, which inhibit the growth of the gram-positive cocci normally found in feces. The incorporation of neutral red into the medium allows lactose fermenters to be differentiated by the pink color of the colonies. Glucose is not present in deoxycholate agar.

181. The answer is C (2, 4). *(Jawetz, ed 13. p 177.)* Although no reliable criteria exist, *Staphylococcus aureus* is considered pathogenic, at least in a laboratory environment, if it produces coagulase or hemolyzes blood. Criteria of even less reliability include the production of yellow pigment, fermentation of mannitol, and liquefaction of gelatin. *S. epidermidis*, which generally is non-pathogenic, is coagulase-negative and nonhemolytic and does not produce a yellow pigment or ferment mannitol. Recently, a novobiocin-resistant biotype of *S. epidermidis*, called *S. saprophyticus*, has been shown to cause disease.

182. The answer is D (4). *(Jawetz, ed 13. p 189.)* The presence of pili in *Neisseria gonorrhoeae* appears to correlate with its virulence. *N. gonorrhoeae* types 1 and 2 have pili and are virulent; the pili aid in the attachment of the organisms to epithelial cells and, in the process, confer protection against phago-cytosis. Nonpiliated organisms are usually nonvirulent.

183. The answer is A (1, 2, 3). *(Fraser, N Engl J Med 297 [1977] :1189-1197. McDade, N Engl J Med 297 [1977] :1197-1203.)* Present methods for the detection of *Legionella pneumophila* in pleural fluid include both microscopic and microbiologic techniques. The organism can be cultured on enriched Mueller-Hinton agar, but only with difficulty, and can be visualized in tissues and body fluids either by direct immunofluorescent staining using conjugated rabbit antisera or by various modifications of the Giemsa stain. *L. pneumophila* resists Gram staining, especially during isolation from tissues and body fluids.

184. The answer is A (1, 2, 3). *(Davis, ed 2. pp 804, 815, 852, 892.)* Recurrent fever is a clinical sign commonly associated with several bacterial illnesses. For example, infection with *Brucella melitensis* (undulent fever) or with *Borrelia recurrentis* (relapsing fever) produces intermittent fever. Tuberculosis can recur chronically. Plague, which is caused by *Yersinia pestis*, is not associated with recurrent fever.

185. The answer is C (2, 4). *(Jawetz, ed 13. pp 215-219.)* *Salmonella-Shigella* (SS) agar is a medium that is useful for the isolation of these organisms from heavily contaminated specimens (e.g., feces). It contains bile salts to inhibit the growth of most gram-positive organisms and citrate to inhibit the growth of coliforms. Bismuth sulfite agar is another medium useful for the isolation of salmonellae. Cultures of salmonellae grown on this medium are characteristically black in color.

186. The answer is C (2, 4). *(Jawetz, ed 13. pp 187-190.)* The neisseriae are strict aerobes; they require the presence of atmospheric oxygen. They grow poorly on media containing salts or fatty acids, while their growth is enhanced by high carbon dioxide tension. For these reasons, cultures taken for neisseriae should be placed in a candle jar and incubated immediately at 37°C on protein-enriched media.

187. The answer is E (all). *(Jawetz, ed 13. pp 182-184.)* Erysipelas and impetigo are streptococcal skin infections most likely to occur in children. Both of these infections can lead to acute glomerulonephritis, although they do so in a very small percentage of cases (0.5 percent). Puerperal fever is a postpartum infection of the uterus.

188. The answer is E (all). *(Jawetz, ed 13. pp 260-261.)* The normal skin flora contains a wide array of microorganisms. Among the bacteria are nonhemolytic staphylococci and diphtheroid bacilli, both of which can be aerobic or anaerobic; alpha-hemolytic streptococci; enterococci; Enterobacteriaceae; and gram-positive, aerobic bacilli. Mycobacteria, fungi, and yeasts also populate the skin of healthy individuals. Resident microorganisms prevent the easy colonization of skin by pathogens.

189. The answer is D (4). *(Jawetz, ed 13. p 262.)* Shortly after birth, the vagina is acidic and is colonized by lactobacilli. As girls become older, the vagina becomes more alkaline, resulting in the appearance of numerous cocci and both gram-positive and gram-negative bacilli. Lactobacilli repopulate the vagina at puberty, when vaginal pH is again acidic.

190. The answer is A (1, 2, 3). *(Davis, ed 2. pp 806-809.)* *Francisella tularensis* is a short, gram-negative organism that is markedly pleomorphic; it is nonmotile and cannot form spores. It has a rigid growth requirement for cysteine. Human tularemia usually is acquired from direct contact with tissues of infected rabbits, but it also can be transmitted by the biting flies and ticks. *F. tularensis* causes a variety of clinical syndromes, including ulceroglandular, oculoglandular, pneumonic, and typhoidal forms of tularemia.

191. The answer is A (1, 2, 3). *(Davis, ed 2. pp 772-774. Volk, p 297.)* Current microbiologic practice delineates only three species of Salmonella: *S. enteritidis*, *S. choleraesuis*, and *S. typhi*. Neither *S. choleraesuis* nor *S. typhi* contains serotypes; *S. enteritidis*, however, contains hundreds of serotypes, including the organisms classically known as *S. derby* and *S. paratyphi*. Salmonellae cause typhoid fever, gastroenteritis, and septicemia.

192. The answer is D (4). *(Davis, ed 2. pp 820-826.)* A clinician should know that both pathogenic and nonpathogenic species of *Bacillus* exist. The distinctive bacteriologic differences between the highly pathogenic *B. anthracis* and the other species, which are essentially nonpathogenic, include the organism's polypeptide capsule and lack of motility. Diseases caused by *B. anthracis* include anthrax and wool-sorter's disease.

193. The answer is A (1, 2, 3). *(Jawetz, ed 13. pp 176, 180, 194, 199.)* The elaboration of potent protein toxin is responsible for the clinical manifestations of diphtheria, tetanus, and many staphylococcal infections. Some clostridia, streptococci, pasteurellae, bordetellae, and shigellae also produce exotoxin. *Mycobacterium tuberculosis* is an invasive organism whose pathogenicity is related to bacterial multiplication and host immune defenses, not to exotoxin production.

194. The answer is A (1, 2, 3). *(Jawetz, ed 13. pp 175, 179, 193, 212.)* Flagella are responsible for the motility of bacteria. Most clostridia possess flagella and therefore are motile. Streptococci, staphylococci, and klebsiellae lack flagellar structures and are nonmotile.

195. The answer is A (1, 2, 3). *(Volk, pp 279-281.)* Final identification of *Neisseria meningitidis* must include carbohydrate utilization tests as well as Gram stain and a test for oxidase. Failure of an organism to agglutinate groups A, B, and C antisera is not uncommon. Isolates of X, Y, and Z strains often are cultured from armed-services personnel.

196. The answer is E (all). *(Jawetz, ed 13. pp 209-210.)* Endotoxins of gram-negative bacteria are heat-stable lipopolysaccharides derived from the cell wall. They are responsible for many of the symptoms and complications of gram-negative infections, including fever, disseminated intravascular coagulation (DIC), leukopenia, and hemorrhagic tissue necrosis. The Shwartzman phenomenon is a complex reaction to experimentally injected endotoxin; it resembles DIC.

197. The answer is D (4). *(Jawetz, ed 13. p 232.)* Chancroid is a venereal disease caused by *Haemophilus ducreyi*, a small gram-negative rod. Meningococci and gonococci, which are gram-negative cocci, cause meningococcal (neisserial) meningitis and ophthalmia neonatorum, respectively. Ludwig's angina is an infection of the oral cavity, in the region of the submaxillary gland; it usually is caused by streptococci.

198. The answer is B (1, 3). *(Jawetz, ed 13. pp 236-237.)* Relapsing fever is characterized by the sudden onset of chills, fever, and headache after an incubation period of three to ten days. Splenomegaly and jaundice often occur. The fever ends abruptly in 3 to 4 days but usually recurs 2 to 14 days later; as many as 10 febrile episodes, which become progressively less severe, can occur. Relapses are thought to be due to alterations in the antigenic structure of *Borrelia recurrentis*, the causative organism. Diarrhea is not a common symptom of relapsing fever.

199. The answer is A (1, 2, 3). *(Davis, ed 2. pp 704-705.)* Pneumococci (*Streptococcus pneumoniae*) and other alpha-hemolytic streptococci often are confused on Gram stain. Pneumococci are generally bile-soluble, sensitive to optochin, and virulent for mice. Both streptococci and pneumococci are catalase-negative and can produce alpha-hemolysis on blood agar.

200. The answer is D (4). *(Davis, ed 2. pp 633-634, 683.)* The capsules of *Streptococcus pneumoniae, Bacillus anthracis,* and *Klebsiella pneumoniae* play a role in the pathogenicity of the organisms. These surface structures inhibit phagocytosis, perhaps by preventing attachment of the leukocyte pseudopod. *Corynebacterium diphtheriae* is nonencapsulated; its pathogenicity is dependent on a protein toxin.

201. The answer is A (1, 2, 3). *(Youmans, pp 133, 642, 775.)* *Bacteroides fragilis* is a constituent of normal intestinal flora. These anaerobic gram-negative rods are uniformly resistant to aminoglycosides and usually to penicillin as well. Reliable laboratory identification may require multiple analytical techniques, including gas chromatographic analysis of metabolic by-products and study of biochemical activity.

202. The answer is B (1, 3). *(Jawetz, ed 13. pp 176-179.)* The spectrum of staphylococcal disease ranges from minor skin infections (carbuncles and impetigo) to life-threatening meningitis and pneumonia. Pertussis, or "whooping cough," is caused by *Bordetella pertussis.* Scarlet fever is a manifestation of beta-hemolytic group A streptococcal infection.

203. The answer is E (all). *(Jawetz, ed 13. p 228.)* Vaccines against *Yersinia pestis* and *Francisella tularensis* may be prepared from avirulent live bacteria, heat-killed or formalin-inactivated virulent bacteria, and chemical fractions of the bacilli. These vaccines provide some immunity to tularemia and bubonic plague, but not to pneumonic plague. Prophylactic tetracycline therapy provides adequate protection to individuals living in areas endemic for plague.

204. The answer is A (1, 2, 3). *(Lennette, ed 2. p 202.)* Klebsiella pneumoniae, an encapsulated gram-negative rod, is a common cause of acute pyogenic pneumonia in alcoholic individuals. The organism is characterized by resistance to ampicillin and carbenicillin, lack of indole production, and ability to use citrate as a sole carbon source. Affected individuals can be treated with kanamycin, gentamicin, and other antibiotics.

205. The answer is E (all). *(Jawetz, ed 13. p 257.)* Listeria monocytogenes is a ubiquitous organism that can be isolated from a wide variety of animals and birds. Infections with Listeria are observed most commonly in immunologically compromised hosts and in patients at the extremes of age. Although Listeria is a short gram-positive rod, it often appears pleomorphic in clinical specimens; this motile, catalase-positive organism produces beta hemolysis on blood agar plates.

206. The answer is C (2, 4). *(Jawetz, ed 13. p 218.)* The species of Shigella can be distinguished as follows:

Species of Shigella	Serotype	Mannitol Fermentation	Decarboxylation of Ornithine
S. dysenteriae	A	−	−
S. flexneri	B	+	−
S. boydii	C	+	−
S. sonnei	D	+	+

Shigellae all ferment glucose. These facultative anaerobic organisms grow best under aerobic conditions.

207. The answer is D (4). *(Finegold, ed 5. pp 217-219, 237-238.)* Both Nocardia and Actinomyces are actinomycetes, a group of filamentous bacteria resembling fungi. Nocardia is an aerobic, branched, gram-positive, rod-shaped bacterium that is only sometimes capnophilic; Actinomyces is an anaerobic, branched, gram-positive rod-shaped bacterium. Both genera are pathogenic: Nocardia causes a pulmonary infection resembling tuberculosis and Actinomyces causes cervicofacial osteomyelitic disease.

208-212. The answers are: 208-C, 209-D, 210-B, 211-A, 212-E. *(Jawetz, ed 13. pp 128-130.)* All five drugs listed in the question are current or former treatments for individuals who have tuberculosis. Cycloserine (D-4-amino-3-isoxazolidinone) acts by inhibiting the incorporation of D-alanine into bacterial cell walls. It is an occasional treatment for urinary tract infections. Cycloserine can cause neurotoxic side-effects and shock.

Patients taking isoniazid (INH) excrete pyridoxine in excess amounts, which leads to peripheral neuritis. The administration of pyridoxine (0.3 to 0.5 g daily) prevents this complication without interfering with the antituberculous effect of INH.

Oral dosage of para-aminosalicylic acid (PAS) for the treatment of individuals suffering from tuberculosis is 8 to 12 g daily. Because this dosage commonly is associated with severe gastrointestinal distress, PAS now is used only rarely.

Streptomycin is extremely toxic for the vestibular portion of cranial nerve VIII, causing ataxia, vertigo, and tinnitus. Although infrequent, hypersensitivity to ethambutol can occur, most commonly manifesting as a visual disturbance.

213-217. The answers are: 213-C, 214-D, 215-E, 216-B, 217-A. *(Jawetz, ed 13. pp 177, 187, 198, 220-221.)* The medium of choice for the isolation of pathogenic neisseriae is Thayer-Martin (TM) agar. TM agar is both a selective and an enriched medium; it contains hemoglobin, the supplement Isovitalex, and the antibiotics vancomycin, colistin, nystatin, and trimethoprim. *Vibrio cholerae* as well as other vibrios, including *V. parahaemolyticus* and *V. alginolyticus*, are isolated best on thiosulfate-citrate bile salts medium, although media such as mannitol-salt agar also support the growth of vibrios. Löwenstein-Jensen slants or plates, which are composed of a nutrient base and egg yolk, are used routinely for the initial isolation of mycobacteria. Löffler's medium, a very rich medium, supports the growth of *Corynebacterium diphtheriae* but suppresses the growth of most other nasopharyngeal microflora. *Staphylococcus aureus* grows very well on sheep blood agar, an agar made up of a nutrient base and 5 to 8% sheep blood; selective and differential media, such as mannitol-salt agar, also are available for *S. aureus.*

218-221. The answers are: 218-E, 219-C, 220-A, 221-D. *(Jawetz, ed 13. pp 223-227.)* The organisms described in the question all are short, ovoid, gram-negative rods. For the most part, they are nutritionally fastidious, requiring blood or blood products for growth. These and related organisms are unique among bacteria in that though they have an animal reservoir they can be transmitted to humans. Humans become infected by a variety of routes, including ingestion of contaminated animal products (*Brucella abortus* in cattle), direct contact with contaminated animal material or with infected animals themselves (*Yersinia enterocolitica* and *Bordetella bronchiseptica* in dogs), and animal bites (*Pasteurella multocida* in many different animals). The laboratory differentiation of these microbes may be difficult and must rely on a number of parameters, including biochemical and serologic reactions, development of specific antibody response in affected individuals, and epidemiologic evidence of infection.

222-225. The answers are: 222-D, 223-A, 224-E, 225-C. *(Finegold, ed 5. pp 134-136, 237-238.)* The organisms listed in the question—*Streptococcus salivarius, S. mutans, Actinomyces viscosus,* and *A. israelii*—all are part of the normal microbiota of the human mouth. Both streptococci are usually alphahemolytic, although nonhemolytic variants may appear, and both are common causes of bacterial endocarditis. *S. mutans* is highly cariogenic (i.e., capable of producing dental caries), in large part because of its unique ability to synthesize a dextran bioadhesive that sticks to teeth. *S. salivarius* settles onto the mucosal epithelial surfaces of the human mouth soon after birth and is often found in the saliva.

Actinomyces organisms are opportunistic members of the normal oral microbiota. Both *A. israelii* and *A. viscosus* are pathogenic and can cause osteomyelitis in the cervicofacial region. Of the two species, *A. israelii*, which is anaerobic, is the more common etiologic agent of actinomycoses. *A. viscosus,* a facultative anaerobe, appears to be cariogenic.

Physiology

DIRECTIONS: Each question below contains five suggested answers. Choose the **one best** response to each question.

226. Penicillinase isolated from *Staphylococcus aureus* inactivates 6-aminopenicillanic acid (shown below) by breaking which of the following numbered bonds?

6-Aminopenicillanic acid

(A) 1
(B) 2
(C) 3
(D) 4
(E) 5

227. The serum of a patient is tested and found to have antibodies to ribitol teichoic acid. The patient most likely has been infected with

(A) *Candida albicans*
(B) *Escherichia coli*
(C) *Klebsiella pneumoniae*
(D) *Staphylococcus aureus*
(E) *Staphylococcus epidermidis*

228. Dinitrophenol kills micro-organisms by

(A) protein coagulation
(B) cell-wall disruption
(C) removal of free sulfhydryl groups
(D) antagonism of oxidative phos-phorylation
(E) poisoning of respiratory enzymes

229. When an agent is introduced into a growing bacterial colony, cell multi-plication ceases; removal of the agent, however, allows bacterial cell division to resume. This agent would be described as

(A) a disinfectant
(B) a bactericide
(C) an antiseptic
(D) a bacteriostat
(E) a sterilizer

230. The internal osmotic pressure of bacteria is between 5 and 20 atmos-pheres. The substance that is responsi-ble for the strength of bacterial cell walls is a

(A) lipoprotein
(B) lipophosphate
(C) liposaccharide
(D) polysaccharide
(E) mucopeptide

231. In the photomicrograph below, the *Escherichia coli* shown in cross section are

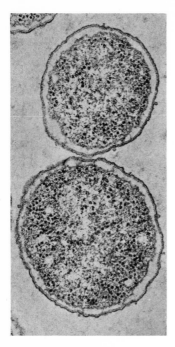

(A) normal in appearance
(B) organizing for mitosis
(C) in a hypotonic environment
(D) partially plasmolyzed
(E) sporulating

232. An aliquot of *Escherichia coli* is treated with ethylenediamine-tetraacetic acid (EDTA). The first wash is analyzed and found to contain alkaline phosphatase, DNase, and penicillinase. The anatomic area of the cell affected by the EDTA is most likely to have been the

(A) periplasmic space
(B) plasma membrane
(C) chromosome
(D) mesosomal space
(E) slime layer

233. Swarming, a characteristic of *Proteus mirabilis*, can be described by which of the following statements?

(A) It requires direct sunlight
(B) It is inhibited on 5% agar
(C) It is potentiated by chloral hydrate
(D) It is a characteristic of all enteric bacteria
(E) It makes isolation of *Proteus* easier

234. A strain of *Staphylococcus aureus*, after exposure to a mutagenic agent, is observed to be devoid of division septa. Such a phenomenon most likely would be caused by a deleterious effect on which of the following organelles?

(A) Phagosome
(B) Lysosome
(C) Ribosome
(D) Mesosome
(E) Somatosome

235. The long bacterial structure shown in the electron micrograph below is necessary for

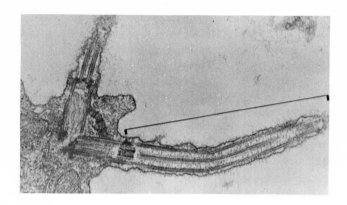

(A) motility
(B) cellular rigidity
(C) active transport
(D) cellular attachment
(E) conjugation

---

236. Actively growing bacteria exposed to penicillin are lysed because penicillin interferes with

(A) conjugation
(B) cell-wall synthesis
(C) amino acid metabolism
(D) nucleic acid metabolism
(E) protein synthesis

237. Analysis of the metabolites produced by an organism's fermentation of glucose reveals small amounts of 6-phosphogluconic acid. This fermentative organism is most likely to be

(A) *Enterobacter*
(B) *Escherichia*
(C) *Leuconostoc*
(D) *Streptococcus lactis*
(E) *Streptococcus faecalis*

238. Bacterial flagella are composed primarily of

(A) carbohydrate
(B) protein
(C) dipicolinic acid
(D) glycoprotein
(E) nucleic acid

239. A recently hired laboratory technologist forgets the iodine-fixation step while Gram-staining a strain of *Staphylococcus*. The most likely result is that the organisms would

(A) appear pink
(B) appear blue
(C) be colorless
(D) wash off the slide
(E) lyse

240. An aerobic organism is incubated in the presence of acetic acid, which is used as a carbon and energy source. Analysis of the metabolic intermediates reveals, among other substances, succinic acid, acetyl CoA, but no pyruvate. The series of reactions responsible for such a pattern of metabolites is called the

(A) succinate cycle
(B) Krebs cycle
(C) TCA cycle
(D) Entner-Doudoroff pathway
(E) glyoxylate cycle

241. Which of the following aqueous ethanol solutions is most bactericidal?

(A) 20%
(B) 40%
(C) 60%
(D) 80%
(E) 100%

242. A chloroform-treated culture of *Mycobacterium tuberculosis* is stained by an acid-fast procedure and then counterstained with methylene blue. The technologist would see

(A) blue organisms against a red background
(B) red organisms against a blue background
(C) blue organisms against a blue background
(D) red organisms against a red background
(E) colorless organisms

243. An unknown isolate is recognized serologically as *Salmonella enteritidis* serotype newport. A mutant of this organism has lost Region 1 (O-specific polysaccharide) of its lipopolysaccharide; this mutant would be identified as

(A) *Salmonella typhi*
(B) *Salmonella newport*
(C) *Salmonella enteritidis*
(D) *Salmonella enteritidis* serotype newport
(E) *Arizona*

244. Protoplasts, spheroplasts, and L forms of bacteria have morphologic and colonial similarities despite the fact that they are taxonomically unrelated. Their morphologic and colonial similarities are related to the

(A) absence of a rigid cell wall
(B) absence of a polysaccharide capsule
(C) presence of a phospholipid outer membrane
(D) presence of endospores
(E) presence of peritrichous flagella

245. In the selection of radiation-induced reversions from a methionine auxotroph, a trace of methionine is required in agar plates of "methionine-free" medium to allow growth of the prototrophic mutant. Requiring an essential factor to produce reversion mutants defines the phenomenon of

(A) induction
(B) mutant resistance
(C) phase variation
(D) phenotypic lag
(E) periodic selection

246. The phenomenon by which a traumatic agent (heat or low pH, for example) converts a bacterial spore into a vegetative cell in a favorable medium is called

(A) activation
(B) emergence
(C) germination
(D) initiation
(E) outgrowth

247. Plasmids are small extrachromosomal genetic elements. Plasmids include all of the following EXCEPT

(A) colicin factors
(B) staphylococcal penicillinase factors
(C) resistance factors
(D) sex factors
(E) transfer factor

248. Ideally, an antibiotic should focus on a microbial target not found in mammalian cells. By this standard, which of the following antibiotic agents would be expected to be most toxic to humans?

(A) Penicillin
(B) Mitomycin
(C) Cephalosporin
(D) Bacitracin
(E) Vancomycin

249. A freeze-fractured *Escherichia coli* is shown below. The elliptical structure at the left is the

(A) plasma membrane
(B) cell wall
(C) cell capsule
(D) cytoplasm
(E) flagellum

DIRECTIONS: Each question below contains four suggested answers of which **one** or **more** is correct. Choose the answer:

A	if	**1, 2, and 3**	are correct
B	if	**1 and 3**	are correct
C	if	**2 and 4**	are correct
D	if	**4**	is correct
E	if	**1, 2, 3, and 4**	are correct

250. Gram-positive bacteria differ from gram-negative bacteria in that gram-positive bacteria

(1) have thicker cell walls
(2) produce endotoxins
(3) are more resistant to drying
(4) do not possess capsules

251. Freeze-etching is a method of preparing cells for electron microscopy. Freeze-etch particles can be located in the

(1) cytoplasm
(2) cell wall
(3) nucleus
(4) cell membrane

252. Bacterial cells in nature possess a glycocalyx. This glycocalyx can be described by which of the following statements?

(1) It commonly is found inside the cell-membrane bilayer
(2) It contains carbohydrate, lipid, and protein moieties
(3) It is a structure unique to bacteria
(4) It determines cell surface properties

253. Gram-positive bacteria lack a periplasmic space. In gram-negative bacteria, the periplasm contains

(1) hydrolytic enzymes
(2) phosphatases
(3) binding proteins
(4) DNA

254. Bacteria that produce propionic acid by propionic fermentation are

(1) found only in the genus *Propionibacterium*
(2) able to extract more energy from substrate than are ethanolic fermenters
(3) important in the manufacture of champagne
(4) slow-growing fermenters of pyruvate

255. Endospores of *Bacillus subtilis* are characterized by

(1) a lack of metabolic activity
(2) greater resistance to drying than the vegetative cell
(3) multiple covering layers, including a peptidoglycan-containing spore-wall cortex
(4) a high calcium content

256. Which of the following statements about the *Escherichia coli* cells shown in the figure below are true?

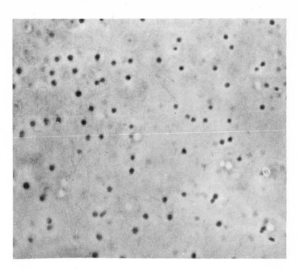

(1) They can result from treatment with penicillin
(2) They can result from treatment with sodium ethylenediamine-tetraacetate (EDTA)
(3) They can result from treatment with lysozyme
(4) They are commonly referred to as endospores

257. Gram-negative and gram-positive bacterial cell walls share which of the following characteristics?

(1) Hydrolysis by lysozyme
(2) Peptide cross-links between polysaccharides
(3) A rigid polysaccharide framework
(4) A wide variety of complex lipids

258. The number of cells in a culture at a given time is a function of the

(1) time elapsed since inoculation
(2) temperature
(3) type of culture medium
(4) size of the inoculum

SUMMARY OF DIRECTIONS

A	B	C	D	E
1, 2, 3 only	1, 3 only	2, 4 only	4 only	All are correct

259. Which of the following are considered hazards of the indiscriminate use of antibiotics?

(1) Development of drug resistance in microbial populations
(2) Direct drug toxicity
(3) Masking of serious infection
(4) Changes in the normal microbial flora with subsequent super-infection

260. The formation of ATP is essential for the maintenance of life. In mammalian systems, the number of moles of ATP formed per gram atom of oxygen consumed (the P/O ratio) is 3; in bacteria, however, the P/O ratio may be only 1 or 2. Reasons for the lower P/O ratio in bacteria include

(1) absence of nicotinamide adenine dinucleotide (NAD)
(2) loss of oxidative phosphorylation coupling sites
(3) less dependence on ATP as an energy source
(4) presence of nonphosphorylative bypass reactions

261. Regulation of branched biosynthetic pathways can be effected by

(1) sequential feedback inhibition
(2) concerted and cumulative feedback inhibition
(3) isofunctional enzymes
(4) enzyme induction

262. Which of the following methods for quantitation of bacteria can be used to measure the total number of **viable** cells?

(1) Turbidometric determination
(2) Microscopic chamber count
(3) Total amount of nitrogen
(4) Plate count

263. Which of the following phases of bacterial culture growth have a zero net-growth rate?

(1) Lag phase
(2) Exponential phase
(3) Maximum stationary phase
(4) Decline phase

264. Phagocytosis plays an important role in the host-parasite interaction. Ingestion of microorganisms causes which of the following effects in a phagocytic cell?

(1) Increased oxygen consumption
(2) Increased glycolysis
(3) Degranulation
(4) Decreased RNA turnover

265. An *Escherichia coli* auxotrophic mutant for the biosynthesis of methionine is likely to

(1) be temperature-sensitive
(2) be resistant to penicillin in a methionine-enriched medium
(3) grow in a sulfur-containing, methionine-free medium
(4) become actively mitotic in a methionine-enriched medium

266. Dipicolinic acid, whose structure is shown below, is a key component of

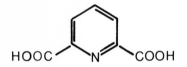

(1) bacterial flagella
(2) eukaryotic cilia
(3) eukaryotic flagella
(4) bacterial spores

267. The F pilus of bacteria is known to

(1) allow cells to attach to a substrate
(2) confer "gender"
(3) function in cell division
(4) be essential for conjugation

268. The "backbone" of peptidoglycan contains

(1) tetrapeptide chains
(2) N-acetylmuramic acid
(3) teichoic acid
(4) N-acetylglucosamine

269. Pasteurization consists of heating a substance at 62°C for 30 minutes. This process can be described by which of the following statements?

(1) It is effective in ridding milk of salmonellal pathogens
(2) It was introduced to sterilize wine
(3) It involves protein denaturation
(4) It reduces the number of viable bacteria by 98 percent

270. Which of the following statements about the sterilization of microorganisms by heat are true?

(1) It is more effective in the presence of water
(2) It is ineffective against most viruses
(3) Vegetative bacteria are likely to be destroyed by boiling for 10 to 15 minutes
(4) Bacterial spores are unlikely to be destroyed by autoclaving at 121°C for 15 minutes

271. Which of the following genes are components of the *lac* operon of *Escherichia coli*?

(1) *y* gene
(2) Promoter (P) gene
(3) *z* gene
(4) Regulator (R) gene

DIRECTIONS: The groups of questions below consist of lettered choices followed by several numbered items. For each numbered item select the **one** lettered choice with which it is **most** closely associated. Each lettered choice may be used once, more than once, or not at all.

Questions 272-275

A 7% sodium dodecyl sulfate polyacrylamide gel electrophoretogram of *Escherichia coli* cell walls is shown below. For each numbered item, choose the lettered band on the electrophoretogram with which it is most likely to be associated.

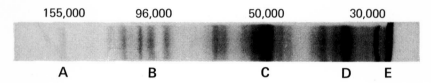

155,000   96,000   50,000   30,000

A   B   C   D   E

272. Lactose permease

273. Beta and beta' RNA polymerase

274. Flagellin

275. Major cell-wall polypeptide

Questions 276-280

For each bacterium listed, choose the lettered form to the right that is most representative of its morphologic structure.

276. *Bacillus subtilis*

277. *Streptococcus pneumoniae*

278. *Staphylococcus aureus*

279. *Streptococcus pyogenes*

280. *Vibrio cholerae*

Questions 281-284

For each numbered item, choose the lettered growth curve (in an exponentially growing culture) with which it is most likely to be associated. (The arrow in the graph indicates the time at which the drugs were added.)

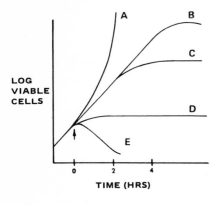

281. Chloramphenicol

282. Penicillin

283. Sulfonamide

284. Control (without antibiotic)

Questions 285-289

For each metabolic reaction listed below, select the bacterium with which it is most characteristically associated.

(A) Mixed acid fermentation
(B) Homolactic fermentation
(C) Yeast alcohol fermentation
(D) Production of 2,3-butanediol
(E) Production of propionic acid

285. *Klebsiella*

286. *Lactobacillus*

287. *Arachnia*

288. *Escherichia coli*

289. *Enterobacter*

Questions 290-292

For each description below, select the process with which it is most likely to be associated.

(A) Conjugation
(B) Recombination
(C) Competence
(D) Transformation
(E) Transduction

290. Uptake by a recipient cell of soluble DNA released from a donor cell

291. Transfer of a donor chromosome fragment by a temperate bacterial virus

292. Direct transfer of a chromosome plasmid between two bacteria

Questions 293-295

For each structure listed, choose the appropriate lettered structure in the freeze-fractured *Escherichia coli* cell shown below.

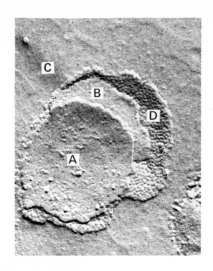

(A) Structure A
(B) Structure B
(C) Structure C
(D) Structure D
(E) None of the above

293. Plasma membrane

294. Eutectic layer

295. Cell wall (lipoid layer)

Physiology

Answers

226. The answer is D. *(Jawetz, ed 13. pp 118-122.)* The structural integrity of the beta-lactam ring in penicillins is essential for their antimicrobial activity. Many resistant strains of staphylococci produce an enzyme, penicillinase, that cleaves the beta-lactam ring at the carbon-nitrogen bond numbered "4" on the diagram accompanying the question. Other organisms, including certain coliform bacteria, produce an amidase enzyme that inactivates penicillin by disrupting bond "1" in the diagram.

227. The answer is D. *(Joklik, ed 16. pp 96-97.)* Teichoic acids are polymers of phosphodiester-bonded polyols. They are linked to the cell walls of gram-positive bacteria by association with muramic acid-6-phosphate. Nearly all strains of *Staphylococcus aureus* produce ribitol teichoic acid, whereas strains of *S. epidermidis* produce glycerol teichoic acid.

228. The answer is D. *(Jawetz, ed 13. pp 108-111.)* Antimicrobial agents interfere with cellular integrity and growth by a variety of actions. These actions include protein coagulation (by heat), cell-wall disruption (penicillin), removal of free sulfhydryl groups (heavy metals), and poisoning of respiratory enzymes (cyanide). Dinitrophenol antagonizes oxidative phosphorylation, inhibiting the energy-yielding process necessary for cell growth and reproduction.

229. The answer is D. *(Jawetz, ed 13. pp 83-84.)* Antimicrobial agents that inhibit bacterial multiplication but do not kill the cells themselves are called bacteriostats. Cell multiplication resumes once the bacteriostatic agent is removed from the environment. Bactericidal agents kill bacteria.

230. The answer is E. *(Jawetz, ed 13. p 13.)* Mucopeptides (also called peptidoglycans and murein) are major components of both gram-positive and gram-negative bacterial cell walls. Mucopeptides are responsible for the tensile strength needed to maintain cell-wall integrity in the face of an internal osmotic pressure of 5 to 20 atmospheres. Lysozymes attack the mucopeptides, thereby weakening the cell wall and resulting in lysis.

231. The answer is D. *(Davis, ed 2. p 130.)* When a bacterial cell, especially a gram-negative cell, is exposed to a hypertonic solution, the cell membrane and contents contract and shrink away from the cell wall. This phenomenon is called plasmolysis. The presence of a rigid cell wall outside the cytoplasmic membrane, as in the photograph presented in the question, is a distinctive feature of plasmolysis.

232. The answer is A. *(Joklik, ed 16. p 40.)* The periplasm is the space between a bacterium's outer membrane and plasma membrane. The periplasmic space in *Escherichia coli* has been shown to contain a number of proteins, sugars, amino acids, and inorganic ions. Ethylenediaminetetraacetic acid (EDTA) is a chelating agent that disrupts the cell walls of gram-negative bacteria.

233. The answer is B. *(Davis, ed 2. p 772.)* The tendency of *Proteus* to swarm — i.e., to spread out rapidly over a solid medium — is the result of active motility. The addition of chloral hydrate to 1 to 2% agar, or the use of 5% agar, inhibits swarming. Swarming makes isolation of *Proteus* difficult.

234. The answer is D. *(Joklik, ed 16. p 43.)* Mesosomes are specialized bacterial organelles often associated with division septa. Mesosomes are formed from invaginations of the cell membrane and aid in the development of cross-walls during cell division. Another characteristic of mesosomes is that they are the site of attachment of the bacterial chromosome.

235. The answer is A. *(Joklik, ed 16. pp 32-34.)* Flagella (as pictured in the photomicrograph accompanying the question) are organelles of motility. They are long, filamentous structures originating in a spherical structure called the basal body. A flagellum is composed of three parts: the filament, the hook, and the basal body.

236. The answer is B. *(Joklik, ed 16. pp 199-201.)* Penicillin acts on actively growing cells. It inhibits the synthesis of bacterial cell walls by blocking the terminal cross-linkage of glycopeptides. Two other antibiotics, vancomycin and bacitracin, also exert their antimicrobial action through impairment of cell-wall synthesis.

237. The answer is C. *(Joklik, ed 16. pp 53-55.)* 6-Phosphogluconic acid is a characteristic metabolic intermediate in the pentose-phosphate metabolic pathway. This pathway is used by heterolactic fermenters such as *Leuconostoc*, the organism responsible for the fermentation of cabbage to form sauerkraut. *Leuconostoc* is a gram-positive bacterium with a dextran capsule.

238. The answer is B. *(Davis, ed 2. p 28.)* Bacterial flagella are composed of flagellin, a protein that has a molecular weight of about 40,000. Chains of flagellin form tightly wound complexes; a triple helical structure is common. Bacterial flagella, unlike protozoal flagella, do not contain microtubules.

239. The answer is A. *(Davis, ed 2. pp 22-23.)* Gram's staining method involves the application of a basic dye, crystal violet, followed by iodine for fixation. The preparation then is treated with ethanol, which decolorizes gram-negative bacteria. Lastly, a red counterstain, safranin, is applied to restain the decolorized organisms. Failure to include the iodine step would prevent the formation of the crystal violet-iodine complex; consequently, the organism would stain gram-negative (pink).

240. The answer is E. *(Joklik, ed 16. p 57.)* In the glyoxylate cycle, acetic acid is oxidized without the formation of pyruvic acid. The net result is the conversion of two acetyl residues to succinic acid. Several enzymatic reactions are common to both the glyoxylate cycle and the tricarboxylic acid (TCA) cycle.

241. The answer is C. *(Davis, ed 2. p 1462.)* Ethanol acts as a disinfectant through protein denaturation, a property enhanced by the presence of water. An aqueous ethanol solution between 50% and 70% would be most effective in sterilization. At ethanol concentrations approaching 100%, or less than 20%, bactericidal action is poor.

242. The answer is C. *(Joklik, ed 16. p 99.)* Treatment of *Mycobacterium tuberculosis* with chloroform removes lipid from the cell wall of the organism. As a result, the organism loses its characteristic "acid-fastness." Chloroform-treated organisms stained by acid-fast procedures appear blue against a blue background if methylene blue is used as the counterstain.

243. The answer is C. *(Joklik, ed 16. p 103.)* Region 1 (the "O-antigenic" side chain of lipopolysaccharide) is responsible for the many serotypes of *Salmonella*. A mutant of *Salmonella* deficient in Region 1 is not identified as a "newport," at least by virtue of its somatic antigen; biochemical identification of this mutant would be *S. enteritidis*. Loss of Region 1 does not affect genus and species classification of *Salmonella*.

244. The answer is A. *(Davis, ed 2. pp 25, 943-944.)* Protoplasts are grampositive bacteria and spheroplasts are gram-negative bacteria that have had their rigid cell walls digested by lysozymes. They must be maintained in a hypertonic medium to prevent lysis. L forms, which are bacteria with defective or absent walls, arise spontaneously under favorable conditions—such as high salt concentrations—or when wall synthesis is impaired by penicillin. Their role in human disease is unclear.

245. The answer is D. *(Davis, ed 2. pp 177-178.)* An auxotrophic organism is unable to survive without provision of a specific nutrient not required by the organism's parent (prototrophic) strain. If a prototrophic reversion is created by treating the auxotrophs with a mutagen (e.g., radiation), supply of the essential nutrient must continue for a brief period of time. During this brief "phenotypic lag," the newly mutated auxotroph grows and divides, allowing segregation of the mutant recessive allele and cytoplasmic expression of the new gene product.

246. The answer is A. *(Davis, ed 2. p 143.)* Although many bacterial spores germinate spontaneously, others cannot be converted into vegetative cells unless they are exposed to a traumatic stimulus. This process is referred to as activation. The effects of aging probably represent the most frequent activating stimuli; heat, solutions of low pH, and sulfhydryl-containing compounds are other recognized traumatic agents.

247. The answer is E. *(Jawetz, ed 13. p 47.)* Plasmids are circular structures made up of double-stranded DNA. Among the common plasmids are the sex factors, the colicin-producing (col) factors, antibiotic resistance (R) factors, and staphylococcal penicillinase plasmids. Transfer factor, a noncellular extract of T lymphocytes, is associated with cell-mediated hypersensitivity; it is functional in delayed hypersensitivity reactions.

248. The answer is B. *(Joklik, ed 16. pp 202-204.)* Antibiotic medications ideally should attack a microbial structure or function not found in human cells. Except for mitomycin, all of the antibiotics listed in the question interfere with cell-wall synthesis in bacteria. Mitomycin inhibits DNA synthesis in both mammalian and microbial systems; viral DNA synthesis, however, is relatively resistant to mitomycin.

249. The answer is A. *(Davis, ed 2. pp 23, 108.)* Freeze-etching involves the freezing of cells at very low temperatures in a block of ice. The ice block is split with a knife, and ice crystals are sublimed — etched — from one of the newly exposed faces. The line of fracture often passes through a natural cleavage plane; in the illustration in the question, for example, are the inner and outer faces of the cell membrane of an *Escherichia coli*. Freeze-etching does not produce the troublesome artifacts introduced during the fixation and drying of specimens.

250. The answer is B (1, 3). *(Davis, ed 2. p 22.)* Because of their thicker cell walls, gram-positive organisms are usually more resistant to drying than gram-negative organisms. Endotoxins are produced only by gram-negative bacteria. Both gram-positive and gram-negative bacteria may possess capsules.

251. The answer is D (4). *(Davis, ed 2. pp 23, 108, 122.)* As a result of freeze-etching, the bacterial cell-membrane bilayer is split, revealing freeze-etch particles.

The particles are thought to represent globular proteins within the hydrophobic region of the membrane. Freeze-etching eliminates many of the alterations induced by other methods of fixation.

252. The answer is C (2, 4). *(Copenhaver, ed 17. pp 51-52.)* On the exterior surface of nearly all animal cells and most bacterial cells is a layer of carbohydrates and associated lipids and proteins. This layer, the glycocalyx, contributes to the surface properties of cells. In bacteria, the glycocalyx is a surface component of the outer membrane.

253. The answer is A (1, 2, 3). *(Davis, ed 2. pp 123-124, 134.)* The periplasm is the substance within the periplasmic space, which is the space between the outer membrane and the cell wall. In addition to various enzymes such as phosphatases and hydrolytic enzymes, the periplasm contains binding proteins, which are involved in membrane transport of certain amino acids and sugars. Grampositive bacteria secrete their hydrolytic enzymes directly into the surrounding environment.

254. The answer is C (2, 4). *(Davis, ed 2. pp 44-45.)* Propionibacterium acnes and Arachnia propionica are capable of fermenting pyruvate to propionic acid and carbon dioxide and generate ATP in the process. This pathway extracts more energy from substrate than is extracted by alcoholic fermentation. Propionic fermentation is important in the commercial manufacture of swiss cheese — carbon dioxide is responsible for the holes, and propionic acid for the odor and flavor.

255. The answer is E (all). *(Davis, ed 2. pp 137-139, 141.)* All spores, including those of *Bacillus subtilis*, exhibit all the characteristics listed in the question. The ability to form spores is a characteristic of three groups of gram-positive organisms: clostridia, bacilli, and sporosarcinae. The function of calcium in spores may be to contract the loose polyanionic cortical peptidoglycan, expelling water and contributing to structural strength. Sporulation is initiated when environmental conditions become unfavorable. Depletion of nitrogen and carbon is thought to play a key role.

256. The answer is B (1, 3). *(Davis, ed 2. p 108.)* The organisms illustrated in the question are spheroplasts of *Escherichia coli*. Spheroplasts are bacteria whose cell walls have been partially removed by the action of lysozyme or penicillin. Ordinarily, with disintegration of the walls the cells undergo lysis; however, in a hypertonic medium the cells persist and assume a spherical configuration. Endospores are formed by several types of gram-positive bacteria.

257. The answer is A (1, 2, 3). *(Joklik, ed 16. p 94.)* A peptidoglycan framework is the basis of both gram-positive and gram-negative bacterial cell walls. This complex network, which imparts rigidity to the cell, is the site of action for lysozyme hydrolysis. (Gram-negative cell walls, it should be noted, are less susceptible to lysozyme than are gram-positive cell walls.) The cell walls of gram-negative organisms are rich in lipopolysaccharides and other complex lipids; gram-positive cell walls, on the other hand, are lipid poor.

258. The answer is E (all). *(Jawetz, ed 13. pp 79-81.)* Microbial growth is affected by many parameters. Environmental temperature, elapsed time in culture, the size of the initial inoculum, and the particular constituents of the medium are all crucial in determining the extent, if any, of microbial growth. Other factors include pH, osmotic pressure, salt concentration, and degree of aeration.

259. The answer is E (all). *(Jawetz, ed 13. p 113.)* All of the negative effects listed in the question can accompany the indiscriminate use of antibiotics. Rational use of antimicrobial drugs demands that a specific etiologic diagnosis be made before antibiotics are administered. Due consideration must be given to every drug's adverse effects.

260. The answer is C (2, 4). *(Davis, ed 2. pp 50-51. Joklik, ed 16. p 59.)* ATP, adenosine triphosphate, is believed to be generated at three reaction points in the electron transport chain: the reductions of flavoprotein, cytochrome b, and cytochrome c. This phenomenon, demonstrated in experiments with mammalian mitochondria, can be expressed in terms of the relationship between the moles of ATP generated for each atom of oxygen consumed—the P/O ratio. In mammalian cells, the P/O ratio is 3.

261. The answer is A (1, 2, 3). *(Jawetz, ed 13. pp 71-72.)* A branched biosynthetic pathway is one in which several substances are formed from a common starting point (e.g., aspartic acid is converted to both lysine and methionine). Regulation of these pathways is by a number of processes. Isofunctional enzymes are different enzymes with the same catalytic activity. One isofunctional enzyme may be inhibited by one amino acid end-product, while a second enzyme is similarly controlled by a different amino acid. Branched pathways also display three types of feedback inhibition: sequential, concerted, and cumulative. Sequential inhibition occurs when two end-products inhibit initial steps in their own branches, causing the accumulation of a substance inhibiting an earlier step common to both branches. In concerted and cumulative feedback inhibition, the enzyme catalyzing the first steps of a branched pathway possesses multiple effector sites, each of which binds a different end-product.

262. The answer is D (4). *(Joklik, ed 16. p 73.)* The only routine method for determining the number of viable bacteria is the standard plate count. Neither turbidometric nor microscopic counts can distinguish between living cells and dead cells. Nitrogen weight relates to total microbial biomass.

263. The answer is B (1, 3). *(Jawetz, ed 13. pp 80-81.)* In the growth of bacterial cultures, the lag phase is a period of synthesis in which the cells prepare for growth by forming new enzymes and intermediates. Growth is maximal during the exponential phase but again reaches a steady state during the maximum stationary phase. Cell death is prominent during the decline phase.

264. The answer is A (1, 2, 3). *(Jawetz, ed 13. p 137.)* When a phagocytic cell "eats" a microorganism, oxygen consumption, glycolysis, and RNA turnover all increase. The phagocytes also undergo degranulation, i.e., their lysosomes rupture to release hydrolytic enzymes into cytoplasmic vacuoles. Phagocytosis can be inhibited by hypophosphatemia, hyperosmolarity, and other conditions.

265. The answer is D (4). *(Davis, ed 2. p 177.)* A methionine auxotrophic mutant cannot synthesize methionine and therefore is unable to grow in a methionine-free medium. Auxotrophs must be grown in media enriched with the essential components (e.g., methionine) that they are unable to produce. Although the growth of mutants may be temperature-sensitive in enriched media, this characteristic is not true of all methionine-requiring auxotrophs. Because penicillin attacks actively multiplying cells, auxotrophs grown in enriched media are susceptible to the drug.

266. The answer is D (4). *(Jawetz, ed 13. pp 24-25.)* Dipicolinic acid, formed in the synthesis of diaminopimelate (DAP), is a prominent component of bacterial spores but is not found in vegetative cells or in eukaryotic appendages. The calcium salt of dipicolinic acid apparently plays an important role in stabilizing spore proteins, but its mechanism of action is unknown. Dipicolinic acid synthetase is an enzyme unique to bacterial spores.

267. The answer is C (2, 4). *(Davis, ed 2. p 195.)* The F sex pilus is a tubular structure possessed only by "male" bacteria. It serves as a conduit in the transfer of genetic information between organisms. Conjugation becomes impossible — and thus "fertility" is lost — if F pili are removed.

268. The answer is C (2, 4). *(Davis, ed 2. p 110.)* Peptidoglycans are components of the cell walls of all bacteria except mycoplasmas and certain halophilic bacteria. They are composed of backbones of N-acetylmuramic acid and N-acetylglucosamine, to which cross-linked tetrapeptides are attached. Attached on occasion to the peptidoglycan backbone are teichoic acids, which are polyolphosphate chains that act as surface antigens.

269. The answer is E (all). *(Davis, ed 2. p 1454.)* Pasteurization, used initially to control the bitterness of wine, is now used primarily to sterilize the pathogens, including salmonellae and streptococci, in milk. In pasteurization, a substance is heated for 30 minutes at 62°C, hot enough to denature many cellular proteins. Total bacterial counts are reduced by 97 to 99 percent.

270. The answer is B (1, 3). *(Davis, ed 2. p 1455.)* Most bacteria, fungi, and viruses are sterilized by boiling for 10 to 15 minutes. However, spores may not be killed by this method. For absolute sterility, autoclaving for 15 minutes at 121°C is recommended. Because proteins are more easily denatured in the presence of water, moist heat is preferable to dry heat.

271. The answer is A (1, 2, 3). *(Davis, ed 2. pp 323-325.)* The *lac* operon of *Escherichia coli* is a group of genes that codes for an inducible enzyme system associated with the ability to use lactose as an energy source. Transcription begins at the promoter (P) gene and, if unhindered, proceeds through the operon's three structural genes: *z*, which codes for β-galactosidase; *y*, which codes for β-galactoside transport protein (lactose permease); and *a*, which codes for β-galactoside transacetylase. Transcription can be turned off if a repressor substance, a product of a regulator (R) gene in another operon, binds to the operator (O) gene located between the P and *z* genes of the *lac* operon.

272-275. The answers are: 272-D, 273-A, 274-C, 275-C. *(Ames, J Biol Chem 249 [1974]:634.)* Gel electrophoresis provides a rapid method for identifying bacterial proteins and estimating molecular weights. A gel can be made of a number of substances, including starch, agar, and polyacrylamide. Starch gel has high separating power, because the fine gel pores act as a molecular sieve. Agar gel is easier to prepare than starch; separation of proteins is accomplished in 30 to 60 minutes. Polyacrylamide gel also separates on the principle of the molecular sieve. It is chemically inert and electrically neutral. The biggest disadvantage of polyacrylamide is that its separating powers are so good that protein patterns, or patterns of other heterogeneous substances, may be too complex to interpret. In the electrophoretogram presented in the question, band A represents RNA polymerases (molecular weight 155,000), band C represents both flagellin and the major cell-wall protein (50,000), and band D represents lactose permease (30,000); band E is the dye front.

276-280. The answers are: 276-D, 277-A, 278-C, 279-B, 280-E. *(Davis, ed 2. p 24.)* The gross shape of a bacterium is both characteristic and important in initial identification. Basically, bacterial shapes fall into one of two categories: coccal (spheroid) and bacillary (rod-like). These classifications can be further subdivided as follows:

- Diplococci, which are paired cocci (e.g., *Streptococcus pneumoniae*);
- Cocci in straight chains (most streptococci, including *S. pyogenes*);
- Cocci in clusters (staphylococci, including *S. aureus*);
- Sarcinae, which are cocci adherent in tetrad or cuboidal arrangements (e.g., *Sarcinia lutea*);
- Simple bacilli (*Bacillus* species, including *B. subtilis*);
- Coccobacilli, which are shorter than most bacilli (e.g., *Neisseria gonorrhoeae*);
- Fusiform bacilli, which are tapered at both ends (*Fusobacterium* species);
- Filamentous bacilli, which are long and thready (*Nocardia* species);
- Vibrios, which are comma-shaped rods (e.g., *Vibrio cholerae*); and
- Spirilla, which are long, spiral-like rods (e.g., *Treponema pallidum*).

281-284. The answers are: 281-D, 282-E, 283-C, 284-B. *(Davis, ed 2. pp 150-152, 160.)* Penicillin causes lysis of growing bacterial cells. Its antimicrobial effect stems from impairment of cell-wall synthesis. Because penicillin is bactericidal, the number of viable cells should fall immediately after introduction of the drug into the medium.

Both chloramphenicol and sulfonamides are bacteriostatic—i.e., they retard cell growth without causing cell death. Chloramphenicol causes an immediate, reversible, bacteriostatic inhibition of protein synthesis. Sulfonamides, on the other hand, compete with para-aminobenzoic acid in the synthesis of folate; intracellular stores of folate are depleted gradually as the cells continue to grow.

The number of viable cells in a culture eventually will level off even if no antibiotic is added to the environment. A key factor in this phenomenon is the limited availability of substrate.

285-289. The answers are: 285-D, 286-B, 287-E, 288-A, 289-D. *(Davis, ed 2. pp 43-46. Joklik, ed 16. pp 54-56.)* Mixed acid fermentation is characteristic of *Escherichia coli*. In this process, substrate is fermented either to lactate or, by the splitting of pyruvate, to formate. *E. coli* also can split formic acid into hydrogen and carbon dioxide in a reaction catalyzed by the enzyme formic hydrogenlyase.

Lactobacillus and most streptococci convert one mole of glucose to two moles of lactic acid in a process known as homolactic fermentation. (In heterolactic fermentation, only one mole of lactate is produced for every mole of glucose.) The souring of milk is a byproduct of lactic fermentation. The production of 2,3-butanediol is characteristic of most *Enterobacter,* *Klebsiella*, and *Serratia* species. Synthesis of this alcohol is the basis for the positive Voges-Proskauer reaction in these organisms. Propionic acid is produced by some gram-positive nonspore-forming rods, such as *Propionibacterium* and *Arachnia*. Propionic acid contributes to the taste and smell of swiss cheese.

290-292. The answers are: 290-D, 291-E, 292-A. *(Jawetz, ed 13. pp 43-53.)* Transformation, transduction, and conjugation are critical processes in which DNA is transferred from one bacterium to another. Transformation, the passage of high-molecular-weight DNA from one bacterium to another, was first observed in pneumococci. Later studies have shown that, at least in *Streptococcus pneumoniae*, double-stranded DNA is "nicked" by a membrane-bound endonuclease, initiating DNA entry into the host cell. One of the nicked DNA strands is digested, and the other is integrated into the host genome.

In conjugation, too, DNA is passed from one bacterium to another. However, instead of the transfer of soluble DNA, a small loop of DNA, called a plasmid, is passed between cells. Examples of plasmids are the sex factors and the resistance (R) factors.

Transduction is a process in which a fragment of donor chromosome is carried to a recipient cell by a temperate virus (bacteriophage). Transduction, which can affect many bacteria, can be "generalized" or "restricted." In generalized transduction, the phage virus can carry any segment of the donor chromosome; in restricted transduction, the phage carries only those chromosomal segments immediately adjacent to the site of prophage attachment.

293-295. The answers are: 293-A, 294-C, 295-D. *(Davis, ed 2. p 23. Van Gool, J Bacteriol 108 [1971]:474-481.)* Freeze fracture is a process in which cells frozen at −150°C are cleaved with a knife. Ice is sublimed from the cleaved surface, and underlying structures are laid bare. The fracture lines in the ice often pass through cells along natural lines of cleavage, revealing internal surfaces through shadowing on microscopy. Natural bacterial cell cleavage planes occur between the peptidoglycan layer and the plasma membrane and between the inner and outer faces of the membrane. In the freeze-fracture photograph presented, the concave fractures from the inside of the envelope out include the plasma membrane (**A**), peptidoglycan layer (**B**), and the lipopolysaccharide layer (**D**); structure **C** is the eutectic layer.

Rickettsiae, Chlamydiae, and Mycoplasmas

DIRECTIONS: Each question below contains five suggested answers. Choose the **one best** response to each question.

296. The rickettsiae are related most closely to

(A) yeast
(B) fungi
(C) viruses
(D) bacteria
(E) protozoa

297. Which of the following rickettsial diseases is transmitted by mites?

(A) Trench fever
(B) Rocky Mountain spotted fever
(C) Tsutsugamushi fever
(D) Epidemic typhus
(E) Endemic typhus

298. A man with chills, fever, and headache is thought to have "atypical" pneumonia. History reveals that he raises chickens and that approximately two weeks ago he lost a large number of them to an undiagnosed disease. The most likely diagnosis of this man's condition is

(A) anthrax
(B) Q fever
(C) relapsing fever
(D) leptospirosis
(E) psittacosis

299. During the course of endemic typhus, affected patients develop antibodies to certain strains of *Proteus vulgaris*. Which of the following statements accurately explains this fact?

(A) This relationship is an example of the Danysz phenomenon
(B) Endemic typhus is caused by a strain of *Proteus vulgaris*
(C) *"Proteus vulgaris"* is the old name for the causative agent, now known to be rickettsial
(D) The causative rickettsiae share antigenic similarities with *Proteus* organisms
(E) Most typhus patients have accompanying *Proteus* bacteremia

300. Which of the following rickettsial diseases is acquired primarily by inhalation?

(A) Scrub typhus
(B) Rickettsialpox
(C) Brill-Zinsser disease
(D) Q fever
(E) Rocky Mountain spotted fever

301. For which of the following rickettsial diseases would the Weil-Felix agglutination reaction be negative?

(A) Rocky Mountain spotted fever
(B) Epidemic typhus
(C) Endemic typhus
(D) Scrub typhus
(E) Q fever

302. A woman who has developed a rash appears in the emergency room of her local hospital. She has a high fever and gives a history of camping in a tick-infested area. Initial screening of the woman's serum reveals a negative Weil-Felix reaction. Which of the following statements can be made about this patient?

(A) She does not have rickettsial disease
(B) She does not have Rocky Mountain spotted fever
(C) She should have a complement-fixation test for rickettsial antibodies
(D) She most likely has Q fever
(E) She most likely has tick paralysis, which can be confirmed by a convalescent Weil-Felix titer

303. Rickettsial growth is enhanced by

(A) chloramphenicol
(B) para-aminobenzoic acid
(C) doxycycline
(D) sulfonamide
(E) tetracycline

304. An inhibitor of ATP synthesis would be expected to retard most severely the host-cell penetration of which of the following organisms?

(A) *Chlamydia psittaci*
(B) *Chlamydia trachomatis*
(C) *Ureaplasma urealyticum*
(D) *Rickettsia rickettsii*
(E) *Mycoplasma pneumoniae*

305. Rocky Mountain spotted fever usually is transmitted to humans by

(A) ticks
(B) mites
(C) lice
(D) fleas
(E) mosquitoes

306. Chlamydiae have an unusual three-stage cycle of development. The correct sequence of these events is

(A) penetration of the host cell, synthesis of elementary body progeny, development of an initial body
(B) penetration of the host cell, development of an initial body, synthesis of elementary body progeny
(C) development of an initial body, synthesis of elementary body progeny, penetration of the host cell
(D) synthesis of elementary body progeny, development of an initial body, penetration of the host cell
(E) synthesis of elementary body progeny, penetration of the host cell, development of an initial body

307. Which of the following organisms would be most likely to appear in a urethral culture from a man who has urethritis?

(A) *Coxiella burnetii*
(B) *Mycoplasma hominis*
(C) *Chlamydia trachomatis*
(D) *Chlamydia psittaci*
(E) *Neisseria sicca*

DIRECTIONS: Each question below contains four suggested answers of which **one** or **more** is correct. Choose the answer:

A	if	**1, 2, and 3**	are correct
B	if	**1 and 3**	are correct
C	if	**2 and 4**	are correct
D	if	**4**	is correct
E	if	**1, 2, 3, and 4**	are correct

308. Chlamydiae can be distinguished from viruses by which of the following characteristics?

(1) Growth outside host cells
(2) Independent protein synthesis
(3) Generation of ATP
(4) Antibiotic sensitivity

309. Which of the following statements about lymphogranuloma venereum, a venereal disease caused by chlamydiae, are true?

(1) It has a first stage characterized by small papules that develop approximately two weeks after infection
(2) It has a second stage characterized by inguinal buboes
(3) It has a third stage characterized by elephantiasis and fibrosis
(4) It may cause severe rectal obstruction and fistula formation

310. Mycoplasmas are bacterial cells that

(1) reproduce on artificial media
(2) have a rigid cell wall
(3) are resistant to penicillin
(4) stain well with Gram stain

311. Two species of chlamydiae, *Chlamydia trachomatis* and *Chlamydia psittaci*, currently are recognized. These organisms have which of the following characteristics?

(1) Only in *C. psittaci* is development of inclusions not inhibited by sulfa drugs
(2) Only the intracellular micro-colonies of *C. trachomatis* contain little glycogen
(3) Both species cause disease in humans
(4) In both species compact non-diffuse inclusions can be observed

312. Trachoma is a chlamydial disease that

(1) is best treated with systemic sulfonamides and ophthalmic tetracycline
(2) affects 400 million individuals worldwide
(3) is a form of chronic kerato-conjunctivitis
(4) can occur in animals other than humans

SUMMARY OF DIRECTIONS

A	B	C	D	E
1, 2, 3 only	1, 3 only	2, 4 only	4 only	All are correct

313. Rickettsiae can be described as

(1) obligate intracellular parasites
(2) energy-deficient
(3) having typical bacterial cell walls
(4) the agents of rat-bite fever

314. Q fever is different from all other rickettsial infections because

(1) it is not associated with a skin rash
(2) it is caused by an organism stable outside the host cell
(3) patients' sera do not contain Weil-Felix antibodies
(4) the causative agent is transmitted by rodents

DIRECTIONS: The group of questions below consists of lettered choices followed by several numbered items. For each numbered item select the **one** lettered choice with which it is **most** closely associated. Each lettered choice may be used once, more than once, or not at all.

Questions 315-319

Match the following.

(A) *Mycoplasma hominis*
(B) *Mycoplasma pneumoniae*
(C) T-strain *Mycoplasma*
(D) *Mycoplasma salivarium*
(E) *Mycoplasma orale*

315. Is associated with development of antibodies to *Streptococcus* MG   B

316. Is associated with venereally transmitted urethritis   C

317. Causes hemagglutination of guinea pig erythrocytes   B

318. Causes hemadsorption of human O erythrocytes   B

319. Splits urea   C

Rickettsiae, Chlamydiae, and Mycoplasmas

Answers

296. The answer is D. *(Davis, ed 2. p 901.)* The rickettsiae were once believed to occupy a special taxonomic class between bacteria and viruses. However, they are now considered to be bacteria, because 1) they contain both RNA and DNA; 2) they are affected adversely by antibiotics; 3) they reproduce by binary fission; and 4) they display many metabolic and chemical characteristics of bacteria.

297. The answer is C. *(Joklik, ed 16. p 735.)* Tsutsugamushi fever (scrub typhus) is a rickettsial disease prevalent in the Far East. Transmitted by a trombicular mite, the disease often produces eschar formation where the mite bite occurred. Rash and fever are prominent signs. Although the fever usually subsides in approximately three weeks, total recovery sometimes requires months. Trench fever and epidemic typhus are transmitted by lice, endemic typhus by fleas, and Rocky Mountain spotted fever by ticks.

298. The answer is E. *(Davis, ed 2. p 926.)* Ornithosis (psittacosis) is caused by *Chlamydia psittaci.* Humans usually contract the disease from infected birds kept as pets, from infected poultry, or during employment in poultry dressing plants. Although ornithosis may be asymptomatic in humans, severe pneumonia can develop. Fortunately, the disease is cured easily with tetracycline.

299. The answer is D. *(Jawetz, ed 13. p 242.)* Because of antigenic similarities between rickettsiae and *Proteus* organisms, the sera of individuals with endemic typhus (a rickettsial disease) are able to agglutinate certain strains of *Proteus vulgaris.* This reaction, known as the Weil-Felix reaction, is used clinically in the diagnosis of rickettsial infections. Other serologic tests, as well as the creation of intraperitoneal infections in laboratory animals, also may be used to identify rickettsial disease.

300. The answer is D. *(Davis, ed 2. pp 911-912.)* Most rickettsial diseases are transmitted to humans by way of arthropod vectors. The only exception is Q fever, the causative agent of which is *Coxiella burneti*. This organism is transmitted by inhalation of contaminated dust and aerosols or by ingestion of contaminated milk.

301. The answer is E. *(Davis, ed 2. p 906.)* Antibodies formed during the course of many rickettsial diseases cross-react with the antigens of certain *Proteus* organisms. This phenomenon occurs in epidemic and endemic typhus, Rocky Mountain spotted fever, and scrub typhus. Testing for *Proteus* antigens (the Weil-Felix reaction) usually is negative in cases of rickettsialpox, Q fever, trench fever, and Brill-Zinsser disease.

302. The answer is C. *(Joklik, ed 16. p 725.)* Clinically, the patient described in the question may have rickettsial disease. It is not uncommon for patients with Rocky Mountain spotted fever, a tick-borne disease, to have a negative Weil-Felix reaction. Complement-fixation tests in these patients usually are positive. Tick paralysis, which occurs mainly in children, produces an ascending paralytic reaction.

303. The answer is D. *(Jawetz, ed 13. p 243.)* Because rickettsial growth is enhanced by sulfonamides, drugs of this class consequently are contraindicated in the treatment of rickettsial diseases. Para-aminobenzoic acid, a structural analog of sulfonamide, inhibits the growth of these organisms. Rickettsial infections may be satisfactorily treated with chloramphenicol or tetracycline.

304. The answer is D. *(Joklik, ed 16. p 726.)* Of the organisms listed in the question, only *Rickettsia rickettsii* penetrates host cells by an active process requiring the expenditure of energy (i.e., ATP). Chlamydiae have a complex growth cycle, which is obligately intracellular. Although the precise mode of penetration is not known, it is likely that a vesicle is formed around the chlamydiae, which then are taken into the cell by a mechanism similar to phagocytosis; chlamydia do not synthesize ATP. *Mycoplasma* species are free-living bacteria that do not actively penetrate cells.

305. The answer is A. *(Jawetz, ed 13. pp 243-245.)* *Rickettsia rickettsii*, the etiologic agent of Rocky Mountain spotted fever, is found in the saliva of the wood tick *Dermacentor andersoni* and the dog tick *D. variabilis*. The rickettsiae can be transmitted to humans by way of a tick's bite. Rocky Mountain spotted fever is not limited to the Rocky Mountains; in fact, most reported cases are in the eastern and southeastern regions of the United States.

306. The answer is B. *(Jawetz, ed 13. p 246.)* The developmental cycle of chlamydiae begins with the "elementary body" attaching to and then penetrating the host cell. The elementary body, now in a vacuole bounded by host-cell membrane, becomes an "initial body." Within about 12 hours the initial body has divided to form many small elementary particles encased within an inclusion body in the cytoplasm; these progeny are liberated by host cell rupture.

307. The answer is C. *(Joklik, ed 16. p 750.)* Chlamydia trachomatis has been associated with 50 percent of cases of nongonococcal urethritis in male individuals. The role of chlamydiae in such infections can be demonstrated by recovery of the agent as well as serologic evidence of infection. *C. trachomatis* is also the etiologic agent of trachoma and inclusion conjunctivitis.

308. The answer is C (2, 4). *(Davis, ed 2. pp 916-918.)* Although both chlamydiae and viruses are obligate intracellular parasites and depend on the host cell for ATP and phosphorylated intermediates, they differ in many respects. Unlike viruses, chlamydiae synthesize proteins, are sensitive to antibiotics, and reproduce by fission. Chlamydiae are readily seen under the light microscope and possess bacterial-type cell walls.

309. The answer is E (all). *(Jawetz, ed 13. pp 250-251.)* Lymphogranuloma venereum, caused by *Chlamydia trachomatis*, has three stages: a papule stage, an inguinal bubo stage, and a fibrosis stage. In women, because of the lymphatic drainage to the perirectal nodes, rectal obstruction and fistulas may appear as late sequelae of the disease. Lymphogranuloma venereum is uncommon in the United States; however, it has been diagnosed in a significant number of veterans who served a tour of duty in Vietnam.

310. The answer is B (1, 3). *(Jawetz, ed 13. pp 254-256.)* Mycoplasmas are extremely small, highly pleomorphic organisms that lack cell walls. They can reproduce on artificial media, forming small colonies with a "fried egg" appearance. They stain poorly with Gram stain but well with Giemsa stain. They are resistant to penicillin but sensitive to tetracycline and sulfonamide.

311. The answer is B (1, 3). *(Joklik, ed 16. p 752.)* The two species of chlamydiae, *Chlamydia psittaci* and *C. trachomatis*, can be distinguished clinically. Microcolonies of *C. psittaci* contain little glycogen and thus do not stain well with iodine. Neither sulfa compounds nor cycloserine inhibits inclusions in *C. psittaci*, but both inhibit inclusions in *C. trachomatis*. The inclusions of *C. psittaci* are irregular and diffuse. The DNA base compositions are dissimilar in the two species.

312. The answer is A (1, 2, 3). *(Jawetz, ed 13. pp 251-252.)* Trachoma is probably the most common cause worldwide of blindness, affecting about 400 million persons. It is a chronic keratoconjunctivitis that can be treated successfully with sulfonamides and tetracycline. Relapse of trachoma, which is an infection with *Chlamydia trachomatis*, is common.

313. The answer is A (1, 2, 3). *(Jawetz, ed 13. pp 241-242.)* Rickettsiae are small, nonmotile bacteria that may appear as either short rods or cocci. Their cell walls resemble those of other gram-negative bacteria. Most rickettsiae are obligate intracellular parasites and cannot be grown in cell-free media. Rickettsiae synthesize ATP, but only very inefficiently. (*Rochalimaea quintana*, the agent of trench fever, is an exception: it can be cultivated on blood agar.) Rat-bite fever is caused by a spirochete.

314. The answer is A (1, 2, 3). *(Davis, ed 2. p 911.)* The etiologic agent of Q fever, *Coxiella burnetii*, is atypical of rickettsiae. It is stable outside the host cell and is resistant to drying. Transmission to humans is by inhalation and not by rodent or arthropod vectors. Rash is not a prominent sign, and Weil-Felix antibodies are not found in the sera of affected individuals.

315-319. The answers are: 315-B, 316-C, 317-B, 318-B, 319-C. *(Davis, ed 2. pp 930-942. Joklik, ed 16. pp 759-761.)* Infection with *Mycoplasma pneumoniae* leads to an increase in antibodies to the alpha-hemolytic *Streptococcus* MG. This agglutinin reaction is probably a nonspecific acute-phase phenomenon. *M. pneumoniae* also can be identified in subculture by its ability to agglutinate human O erythrocytes and adsorb guinea pig red blood cells.

The so-called T strains of *Mycoplasma* (now termed *Ureaplasma urealyticum*) have been implicated in cases of nongonococcal urethritis. As the name implies, this organism is able to split urea, a fact of diagnostic significance. *U. urealyticum* is part of the normal flora of the genitourinary tract, particularly in women.

The only other species of *Mycoplasma* strongly associated with human disease is *M. hominis*. A normal inhabitant of the genital tract of women, this organism has been demonstrated to produce an acute respiratory illness that is associated with sore throat and tonsillar exudate but not with fever.

Mycology

DIRECTIONS: Each question below contains five suggested answers. Choose the **one best** response to each question.

320. Although candidiasis of the oral cavity (thrush) usually is controlled by the administration of nystatin, the disseminated or systemic form of candidiasis requires vigorous therapy with

(A) penicillin
(B) amphotericin B
(C) interferon
(D) chloramphenicol
(E) thiabendazole

321. Septate hyphae, 3 to 4 μm in diameter, are seen in hematoxylin-eosin-stained lung tissue. On the basis of this finding, a technologist would most likely identify the fungus present in the sample as

(A) *Rhizopus*
(B) *Aspergillus*
(C) *Mucor*
(D) *Blastomyces dermatitidis*
(E) *Histoplasma capsulatum*

322. All of the following statements about the actinomycetes are true EXCEPT

(A) they are gram-positive
(B) they can occur normally in the oral cavity
(C) they possess branched mycelia
(D) they have cellulose-containing cell walls
(E) they actually are filamentous bacteria

323. The major cause of favus, a severe form of chronic ringworm of the scalp, is

(A) *Trichophyton schoenleinii*
(B) *Trichophyton rubrum*
(C) *Microsporum canis*
(D) *Pityrosporum furfur*
(E) *Epidermophyton floccosum*

324. Which of the following fungi grows primarily within cells of the reticuloendothelial system?

(A) *Sporothrix schenckii*
(B) *Histoplasma capsulatum*
(C) *Cryptococcus neoformans*
(D) *Coccidioides immitis*
(E) *Blastomyces dermatitidis*

325. In the photomicrograph below, the object designated by the arrow is

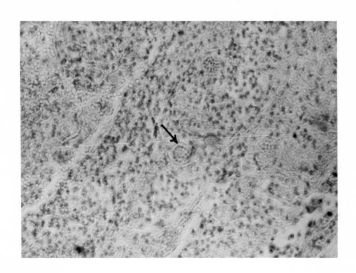

(A) an encapsulated yeast
(B) a thick-walled spore
(C) a spherule
(D) a hyphal strand
(E) a macroconidium

326. The infectious particle in coccidioidomycosis is the

(A) budding yeast
(B) sporangiospore
(C) arthrospore
(D) encapsulated yeast
(E) mature sporangium

327. A diagnostic characteristic of *Candida albicans* grown on corn-meal agar is

(A) production of a red pigment
(B) mucoid colony with a pigmented center
(C) lack of pigment formation
(D) production of chlamydospores
(E) production of dematiaceous hyphae

328. A dimorphous fungus is one that

(A) produces arthrospores and
    chlamydospores
(B) reproduces both sexually and
    asexually
(C) can grow as a yeast or a mold
(D) forms protoplasts
(E) invades hair and skin

329. Direct microscopic examination
of a sputum specimen digested with
10% sodium hydroxide reveals an encapsulated yeast 4 to 20 μm in diameter. This organism is most likely to be

(A) *Candida albicans*
(B) *Cryptococcus neoformans*
(C) *Geotrichum candidum*
(D) *Aspergillus fumigatus*
(E) *Blastomyces dermatitidis*

330. Finding "sulfur granules" in a
wound is most indicative of infection
with

(A) *Nocardia asteroides*
(B) *Candida albicans*
(C) *Cryptococcus neoformans*
(D) *Actinomyces israelii*
(E) *Geotrichum candidum*

331. In the culturing of fungi, which
of the following media should be used
routinely?

(A) Sabouraud's agar
(B) Blood agar
(C) SS agar
(D) Chocolate agar
(E) Thioglycollate broth

332. *Geotrichum* is a fungus that
would be LEAST likely to infect the

(A) alimentary tract
(B) lungs
(C) bronchi
(D) brain
(E) oral cavity

333. During the third trimester of
pregnancy, vaginal infection with
which of the following organisms
occurs more frequently than normal?

(A) *Candida*
(B) *Acinetobacter*
(C) *Aspergillus*
(D) *Microsporum*
(E) *Epidermophyton*

334. Individuals who have disseminated coccidioidomycosis can demonstrate any of the following EXCEPT

(A) a positive coccidioidin skin test
(B) a negative coccidioidin skin test
(C) a high titer of complement-
    fixing antibodies
(D) immunity to reinfection
(E) therapeutic response to
    griseofulvin

335. A section of tissue from a foot of a person assumed to have maduromycosis (fungal mycetoma) shows a lobulated granule composed of fungal hyphae. In the United States, the most common etiologic agent of this condition is a species of

(A) *Scopulariopsis*
(B) *Nocardia*
(C) *Actinomyces*
(D) *Monosporium*
(E) *Epidermophyton*

336. The most accurate diagnosis of fungal disease rests upon the isolation of fungi from lesions. Visualization of fungi in a clinical specimen is best accomplished by treating the specimen with

(A) silver nitrate
(B) hydrochloric acid
(C) potassium hydroxide
(D) para-aminobenzoic acid
(E) griseofulvin

337. There are three genera of dermatophytes: *Epidermophyton, Microsporum,* and *Trichophyton.* Infections caused by these organisms (dermatophytoses) are

(A) marked by alveolar irritation
(B) characterized by aflatoxin-induced hallucinations
(C) confined to keratinized tissues
(D) rarely associated with chronic lesions
(E) easily treatable with penicillin

338. Which of the following fungi is commonly found in fresh bird droppings?

(A) *Histoplasma capsulatum*
(B) *Blastomyces dermatitidis*
(C) *Sporothrix schenckii*
(D) *Cryptococcus neoformans*
(E) None of the above

339. The most common form of sporotrichosis is

(A) lymphatic sporotrichosis
(B) disseminated sporotrichosis
(C) visceral sporotrichosis
(D) pulmonary sporotrichosis
(E) osseous sporotrichosis

DIRECTIONS: Each question below contains four suggested answers of which **one** or **more** is correct. Choose the answer:

A	if	**1, 2, and 3**	are correct
B	if	**1 and 3**	are correct
C	if	**2 and 4**	are correct
D	if	**4**	is correct
E	if	**1, 2, 3, and 4**	are correct

340. Mucormycosis, also called phycomycosis, is commonly caused by *Mucor* and *Rhizopus*. Which of the following statements about this disorder are true?

(1) It occurs largely as a complication of a chronic debilitating disease, such as diabetes mellitus
(2) It may produce thrombosis and infarction of arterioles
(3) It usually begins in the upper respiratory tract
(4) It usually is diagnosed at autopsy

341. *Coccidioides immitis* can be associated with

(1) arthrospores
(2) granulomatous disease
(3) thin-walled cavities in lungs
(4) "valley fever" or "desert rheumatism"

342. Which of the following statements about *Histoplasma capsulatum*, a fungus endemic in the midwestern United States, are true?

(1) It assumes a yeast form when isolated in Sabouraud's glucose agar at 25°C
(2) It primarily causes acute pulmonary infection
(3) Infection may cause a radiologic picture of scattered calcifications in the heart valves
(4) It is found in soil near chicken coops and bat caves

343. Several cases of brain abscess caused by chromoblastomycotic fungi have been reported worldwide. These abscesses are caused most often by species of

(1) *Cladosporium*
(2) *Cryptococcus*
(3) *Fonsecaea*
(4) *Candida*

344. Fungi differ from bacteria in that they

(1) contain no peptidoglycan
(2) have nuclear membranes
(3) are susceptible to griseofulvin
(4) are prokaryotic

345. Cryptococci have a polysaccharide capsule that

(1) is an aid to diagnosis
(2) inhibits phagocytosis
(3) cross-reacts with rheumatoid factor
(4) causes a precipitin reaction with hyperimmune rabbit serum

346. In humans, fungal disease can be produced by

(1) invasion of keratin-rich tissues
(2) contamination of wounds with spores or mycelial fragments
(3) inhalation of spores
(4) invasion of mucous membranes

347. Reproductive mechanisms in fungi include

(1) sporulation followed by spore germination
(2) hyphae fragmentation
(3) budding
(4) binary fission

DIRECTIONS: The groups of questions below consist of lettered choices followed by several numbered items. For each numbered item select the **one** lettered choice with which it is **most** closely associated. Each lettered choice may be used once, more than once, or not at all.

Questions 348-351

For each constitutional disorder, choose the supervening infection with which it is most likely to be associated.

(A) Endocardial candidiasis
(B) Aspergillosis
(C) Mucormycosis
(D) Nocardiosis
(E) Sporotrichosis

348. Diabetes mellitus

349. Bronchiectasis

350. Pulmonary alveolar proteinosis

351. Drug addiction

Questions 352-356

For each skin disease below, select the organism most likely to be the causative agent.

(A) *Epidermophyton floccusum*
(B) *Pityrosporum furfur*
(C) *Microsporum canis*
(D) *Cladosporium werneckii*
(E) *Trichosporon cutaneum*

352. Tinea corporis (ringworm)   C

353. Tinea cruris (jock itch)   A

354. Tinea pedis (athlete's foot)   A

355. Tinea capitis   C

356. Tinea versicolor   B

Mycology

Answers

320.   The answer is B.   *(Jawetz, ed 13. p 277.)* Disseminated candidiasis can be life-threatening either as a primary infection in immunosuppressed patients or as a secondary infection of the lungs, kidneys, and other organs in individuals who have tuberculosis or cancer. These persons must be treated with amphotericin B or other antifungal drugs. Nystatin, the treatment for candidiasis of the mouth (thrush), does not reach tissues and thus is ineffective in treating disseminated candidiasis.

321.   The answer is B.   *(Davis, ed 2. pp 986, 997-998.)* *Rhizopus* and *Mucor* have large (up to 15 µm in diameter) nonseptate hyphae. In contrast, the hyphae of *Aspergillus* are 3 to 4 µm in diameter, are septate, and show dichotomous branching. *Blastomyces dermatitidis* is a thick-walled, multinucleated, spherical cell 8 to 10 µm in diameter, without a capsule when seen on tissue section. *Histoplasma capsulatum* is a small oval yeast cell 1 to 3 µm in diameter; it can be found within macrophages and reticuloendothelial cell of infected tissue.

322.   The answer is D.   *(Jawetz, ed 13. pp 280-281.)* Although the actinomycetes are bacteria, they superficially resemble fungi. Some of these gram-positive organisms are aerobic, some anaerobic. Actinomycetes are characterized by branched mycelia; they possess a muramic acid-containing wall, in contrast to true fungal cell walls, which contain chitin and cellulose. Some genera (e.g., *Actinomyces*) are normal inhabitants of the oral cavity.

323.   The answer is A.   *(Davis, ed 2. pp 1004-1005.)* *Trichophyton schoenleinii* infection may cause favus, which is characterized by destruction of hair follicles and permanent loss of hair. Scutula (cup-like structures) are formed by crusts around the infected follicles. *Pityrosporum furfur* causes a fungal skin infection producing brownish-red scaling patches on the neck, trunk, and arms. *Epidermophyton floccosum* and *T. rubrum* are common causes of athlete's foot. *Microsporum canis* infections involve the hair and skin and can be differentiated from *Trichophyton* infections by the ability of *Microsporum* to fluoresce under ultraviolet light.

324. The answer is B. *(Jawetz, ed 13. p 273.)* Histoplasmosis is an intracellular mycosis of the reticuloendothelial system. Clinical findings include lymphadenopathy, enlarged spleen and liver, high fever, and anemia. The characteristic lesion shows focal areas of necrosis in small granulomas. The small, oval, yeast-like cells are found within phagocytic cells of the infected organ system.

325. The answer is B. *(Davis, ed 2. p 989.)* Thick-walled spores, as shown in the photomicrograph accompanying the question, are characteristic of many fungal infections, including blastomycosis, coccidioidomycosis, and histoplasmosis. Observation of these structures in sputum or in tissue should alert the microbiologist to a diagnosis of systemic fungal infection. The presence of encapsulated yeast in clinical specimens may suggest the presence of *Cryptococcus*.

326. The answer is C. *(Jawetz, ed 13. pp 272-273.)* The source of infection in coccidioidomycosis is the inhaled arthrospore. Affected individuals may have either an asymptomatic respiratory infection or an influenza-like illness. Most persons recover completely with symptomatic treatment and rest. Amphotericin B is the treatment of choice for disseminated coccidioidomycosis.

327. The answer is D. *(Jawetz, ed 13. pp 277-278.)* When grown on corn-meal agar, *Candida albicans* produces both chlamydospores 8 to 12 μm in diameter and clusters of blastospores. Each of these events is diagnostically characteristic of *C. albicans*. Specimens cultured on Sabouraud's glucose agar at room temperature show blastospores and pseudomycelia.

328. The answer is C. *(Davis, ed 2. p 977.)* Many species of fungi, including most fungi pathogenic for humans, can grow as yeasts or molds, depending on the environment. An example of these dimorphic fungi is *Blastomyces dermatitidis*, which grows as a mold at 25°C but as a yeast at 37°C. In infected cells, dimorphic fungi usually appear as yeasts, and, when cultivated in vitro, as molds.

329. The answer is B. *(Jawetz, ed 13. pp 278-279.)* *Cryptococcus neoformans* usually forms a mucoid capsule that can be detected in unstained preparations if a suitable mounting fluid is used. Neither *Candida albicans* nor *Blastomyces dermatitidis* forms capsules. *Aspergillus fumigatus* would be present as hyphae, and *Geotrichum candidum* as hyphae and dissociated cells.

330. The answer is D. *(Davis, ed 2. pp 875-876.)* The finding of yellow "sulfur granules" in an abscess indicates actinomycosis. (Although the presence of these granules facilitates identification of the etiologic agent, it is not necessary for diagnosis.) "Sulfur granules" are named for their yellow color and not for their chemical composition; in fact, these granules are actually small groupings of actinomycetic colonies. Most actinomycotic abscesses are mixed infections, and washed "sulfur granules" may contain colonies of various bacteria, including fusiform bacilli and anaerobic streptococci.

331. The answer is A. *(Jawetz, ed 13. pp 271-279.)* Sabouraud's glucose agar is the medium of choice for culturing most fungi. On Sabouraud's agar, *Candida albicans* incubated at room temperature grows as soft, cream-colored colonies. *Cryptococcus neoformans* is mucoid and cream-colored and has no mycelia. White to brownish filamentous colonies are seen with *Blastomyces dermatitidis* grown at room temperature, and white, cottony colonies made up of tuberculate spores are diagnostic of *Histoplasma capsulatum.*

332. The answer is D. *(Jawetz, ed 13. p 276.)* *Geotrichum candidum* is a yeast-like fungus that may be a normal inhabitant of the mouth and gut. It produces an infection of bronchi, lungs, and mucous membranes known as geotrichosis. Clinical findings include chronic bronchitis and thrush-like lesions of the mouth. Treatment consists of oral administration of potassium iodide and topical gentian violet (1%) for oral lesions.

333. The answer is A. *(Davis, ed 2. pp 995-996.)* *Candida*, and in particular *C. albicans*, may be found as part of the normal flora of the mouth, vagina, and gastrointestinal tract; as a pathogen, it is an opportunistic fungus. When invasive, candidiasis can be an acute or chronic infection, either localized or disseminated. High blood sugar levels in women in the third trimester of pregnancy (as well as in women who have diabetes) encourage vulvovaginal candidiasis.

334. The answer is E. *(Davis, ed 2. p 993. Jawetz, ed 13. pp 271-273.)* In individuals who have coccidioidomycosis, a positive skin test to coccidioidin appears 2 to 21 days after the appearance of disease symptoms and may persist for 20 years without re-exposure to the fungus. A decrease in intensity of the skin response often occurs in clinically healthy people who move away from endemic areas. A negative skin test often is associated with disseminated disease. Complement-fixing immunoglobulin G (IgG) antibodies, which may not appear at all in mild disease, rise to a high titer in disseminated disease, a poor prognostic sign. Most individuals infected with *Coccidioides immitis* are immune to reinfection. Intravenous amphotericin B is the treatment of choice for disseminated coccidioidomycosis.

335. The answer is D. *(Joklik, ed 16. p 1075.)* Maduromycosis is a slowly progressing disease of the subcutaneous tissues and is caused by a variety of fungi. The term "Madura foot" has been used to describe the foot lesion. Although several fungi have been isolated in the United States from individuals who have maduromycosis, *Monosporium apiospermum* appears to be the most common.

336. The answer is C. *(Davis, ed 2. p 981.)* Evidence supporting the presence of fungal infection includes the clinical appearance of lesions and positive serologic reactions. However, detection of fungi in lesions, either by microscopic inspection or culture, is the best evidence. Treating a specimen with 10% sodium hydroxide or potassium hydroxide hydrolyzes protein, fat, and many polysaccharides, leaving the alkali-resistant cell walls of most fungi intact and visible.

337. The answer is C. *(Jawetz, ed 13. p 266.)* The dermatophytes are a group of fungi that infect only superficial keratinized tissue (skin, hair, nails). They form hyphae and arthrospores on the skin; in culture, they develop colonies and spore forms. Tinea pedis, or athlete's foot, is the most common dermatophytosis. Several topical antifungal agents, such as undecylenic acid, salicylic acid, and ammoniated mercury, may be useful in treatment. For serious infection, systemic use of griseofulvin is effective.

338. The answer is E. *(Davis, ed 2. pp 986, 988, 990, 997, 1000.)* Histoplasma *capsulatum* and *Cryptococcus neoformans*, though isolated quite commonly from soil contaminated with bird droppings, are not usually present in fresh droppings. The droppings appear, therefore, to enrich the soil, making it a more favorable culture medium. Birds are highly resistant to both these organisms, which can remain viable in dried material for months. *Sporothrix schenckii* grows in soil and on vegetation. The distribution of *Blastomyces dermatitidis* is not known, although it may be in soil or organic debris.

339. The answer is A. *(Davis, ed 2. pp 999-1000.)* The most common form of sporotrichosis involves the lymphatics draining the primary cutaneous lesion. The classic syndrome is a chronic ulcer on a finger or hand and an associated chain of enlarged lymph nodes extending up to the arm. If untreated, the mycosis may disseminate.

340. The answer is E (all). *(Davis, ed 2. pp 998-1000.)* Inhalation of spores of *Mucor* or *Rhizopus* is most likely to cause infection in individuals who have diabetes mellitus, uremia, or other chronic disorders or who are immunosuppressed. Large hyphae extend through contiguous tissues and produce angiitis, thrombi, and necrosis. Clinical lesions are dark in appearance (brown to black) and commonly are seen in the oropharynx.

341. The answer is E (all). *(Jawetz, ed 13. pp 271-273.)* Coccidioidomycosis is a fungal disease endemic in the southwestern United States. Inhalation of arthrospores can cause either an asymptomatic respiratory infection or an influenza-like illness. Less than 10 percent of affected individuals subsequently develop a hypersensitivity reaction, which manifests as erythema nodosum or erythema multiforme; this disorder, termed "valley fever" or "desert rheumatism," may be associated with thin-walled cavities in the lungs. All symptoms tend to subside spontaneously. In less than one percent of affected individuals, disseminated disease occurs; in these persons, granulomas indistinguishable from those of tuberculosis can be seen in all organs.

342. The answer is D (4). *(Jawetz, ed 13. pp 273-274.)* *Histoplasma capsulatum* causes histoplasmosis and can be isolated from soil enriched by bird feces and bat droppings. Asymptomatic infection is common and can be associated with calcified foci seen in lungs, spleen, and liver on x-ray. In a minority of cases, disseminated disease occurs, with involvement of the reticuloendothelial system. *H. capsulatum* grows in yeast form on blood agar at 37°C: when incubated on Sabouraud's agar at room temperature, diagnostic tuberculate spores develop.

343. The answer is B (1, 3). *(Davis, ed 2. p 1000. Joklik, ed 16. p 1063.)* Although rare, disseminated fungal infections may occur following skin infection with fungi of the family Dematiaceae. These fungi are characterized by black or dark brown spores and most typically cause a disease called chromoblastomycosis, a slowly progressing granulomatous skin disease. Two of these fungi, *Cladosporium trichoides* and *Fonsecaea pedrosoi*, have been reported to cause brain abscess.

344. The answer is A (1, 2, 3). *(Davis, ed 2. p 980.)* Fungi are eukaryotic, while bacteria are prokaryotic. Fungi are sensitive to griseofulvin; bacteria are not. Bacterial cell walls contain peptidoglycan and their nuclear material is not membrane-bound; fungi have neither of these characteristics.

345. The answer is E (all). *(Davis, ed 2. pp 984-985.)* The characteristic capsules of cryptococci allow the yeast cells to be seen easily in India-ink suspensions. All strains of cryptococci produce capsules. Only *Cryptococcus neoformans* is pathogenic in humans, who have a weak immune response to this organism; however, hyperimmunized rabbits produce capsule-specific antisera that differentiate among three strains of *C. neoformans*. The precipitin reaction with rabbit serum is the only serologic test of diagnostic value. Latex tests for cryptococcal antigen may be falsely positive due to a cross-reaction with rheumatoid factor.

346. The answer is E (all). *(Davis, ed 2. p 979.)* Cutaneous mycoses (dermatophytoses) and superficial mycoses cause disease in skin, hair, and nails by invasion of keratinized tissue. Systemic fungal disease is caused by inhalation of spores. Infections due to direct implantation of spores in the skin, called subcutaneous mycoses, are caused by soil saprophytes. Fungi of normal flora can directly invade a susceptible host through mucous membranes and cause local or disseminated disease (vulvovaginal candidiasis in pregnant or diabetic women, for example).

347. The answer is A (1, 2, 3). *(Davis, ed 2. pp 971-972.)* Budding is the major asexual reproductive process in yeasts. Other forms of vegetative reproduction in fungi include fragmentation of hyphae and sporulation followed by germination. These methods yield new clones without nuclear fusion. Sexual reproduction with fusion of donor and recipient cell nuclei allows for genetic variation among the four haploid cells that are formed.

348-351. The answers are: 348-C, 349-B, 350-D, 351-A. *(Davis, ed 2. pp 995-998. Jawetz, ed 13. pp 140-141.)* Certain underlying constitutional diseases or conditions have an increased association with particular supervening infections, including several that are fungal in origin. In addition to lowered host resistance, factors leading to the establishment of a supervening infection include hormonal imbalances and drug influences, notably immunosuppressive therapy. Women who have diabetes mellitus are more susceptible than healthy nondiabetic women to mucormycosis and vulvovaginal candidiasis. *Candida* endocarditis is a rare disorder known to affect drug addicts. Chronic obstructive lung disease causing impaired movement of bronchial secretions (e.g., bronchiectasis and bronchial carcinoma) can predispose affected individuals to such fungal infections as aspergillosis and bronchopulmonary candidiasis. Secondary infection with *Nocardia* has been associated with pulmonary alveolar proteinosis.

352-356. The answers are: 352-C, 353-A, 354-A, 355-C, 356-B. *(Davis, ed 2. pp 1000-1006. Jawetz, ed 13. pp 266-269.)* Dermatomycoses are cutaneous mycoses caused by three genera of fungi: *Microsporum, Trichophyton,* and *Epidermophyton.* These infections are called tinea or ringworm, a misnomer that has persisted from the days when they were thought to be caused by worms or lice. Tinea capitis (ringworm of the scalp) is due to an infection with *M. canis* or *T. tonsurans.* It usually occurs during childhood and heals spontaneously at puberty. Circular areas on the scalp with broken or no hair are characteristic of this disorder. Tinea corporis (ringworm of the body) is caused by *M. canis* and *T. mentagrophytes.* This disorder affects smooth skin, producing circular pruritic areas of redness and scaling. Tinea cruris (ringworm of the groin, "jock itch") and tinea pedis (ringworm of the feet, "athlete's foot") both are caused by *T. rubrum, T. mentagrophytes,* or *E. floccosum.* Both of these common conditions are pruritic and either can cause scaling.

Tinea versicolor (pityriasis versicolor) is not a dermatomycotic condition. Rather, it is a superficial mycosis now thought to be caused by *Pityrosporum furfur* (*Malassezia furfur* traditionally has been labeled the etiologic agent). The disorder is characterized by chronic but asymptomatic scaling on the trunk, arms, or other parts of the body.

Parasitology

DIRECTIONS: Each question below contains five suggested answers. Choose the **one best** response to each question.

357. Malaria is not now indigenous to the United States in part because of

(A) "herd" immunity
(B) mandatory one-month quarantine of all immigrants from endemic regions
(C) elimination of the reservoir of infected culicine mosquitoes
(D) the effectiveness and ready availability of quinine-derived drugs
(E) chlorination of drinking water

358. A teenager who has recently visited a Rocky Mountain ski resort complains of weakness, malaise, abdominal cramps, and diarrhea. Which of the following laboratory results would be most compatible with a diagnosis of giardiasis?

(A) Worms in a warm stool specimen
(B) Cysts in a formed stool specimen
(C) Trophozoites in the urine
(D) Round, thick-walled trophozoites in a fatty formed stool
(E) Parasitic inclusions in red blood cells

359. An organism capable of living either free or as a parasite is called

(A) an obligate parasite
(B) an erratic parasite
(C) an operculate parasite
(D) a pseudoparasite
(E) a facultative parasite

360. Hydatid disease in humans may be caused by the larval or hydatid stage of *Echinococcus granulosus*. Which of the following animals are most likely to be hosts for this parasite?

(A) Dogs
(B) Cows
(C) Horses
(D) Rats
(E) Pigs

361. A diagnosis of trichinosis can be established most definitively by the presence of

(A) larvae in skeletal muscle
(B) eggs in three consecutive fecal specimens
(C) eosinophilia
(D) periorbital edema
(E) myositis

362. The parasite shown in the blood smear pictured below commonly remains in the lymphatic system in the early stages of infestation. This organism, which is the causative agent of elephantiasis, is

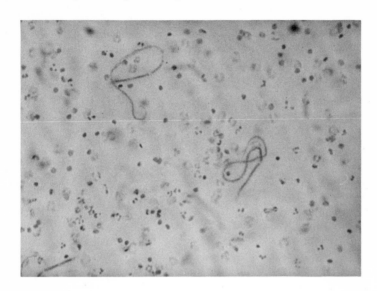

(A) *Strongyloides stercoralis*
(B) *Ancylostoma duodenale*
(C) *Ascaris lumbricoides*
(D) *Wuchereria bancrofti*
(E) *Trichuris trichiura*

363. Autoinfection may be responsible for long-standing disease with

(A) *Paragonimus westermani*
(B) *Necator americanus*
(C) *Strongyloides stercoralis*
(D) *Schistosoma haematobium*
(E) *Diphyllobothrium latum*

364. Which of the following techniques is employed most successfully for recovering pinworm eggs?

(A) Sugar fecal flotation
(B) Zinc-sulfate fecal flotation
(C) Tap-water fecal sedimentation
(D) Direct fecal centrifugal flotation
(E) Anal swabbing with cellophane tape

365. The insect vector of *Trypanosoma cruzi*, the cause of Chagas' disease, is the

(A) rat mite
(B) anopheline mosquito
(C) Lone Star tick
(D) reduviid bug
(E) head louse

366. Visceral larva migrans, an invasion of human viscera by nematodes, usually is produced by

(A) *Enterobius vermicularis*
(B) *Ascaris lumbricoides*
(C) *Toxocara canis*
(D) *Ancylostoma braziliense*
(E) *Ancylostoma duodenale*

367. Human infection with the beef tapeworm *Taenia saginata* usually is less serious than infection with the pork tapeworm *T. solium* because

(A) acute intestinal stoppage is less common in beef tapeworm infection
(B) larval invasion does not occur in beef tapeworm infection
(C) toxic by-products are not given off by the adult beef tapeworm
(D) the adult beef tapeworms are smaller
(E) beef tapeworm eggs cause less irritation of the mucosa of the digestive tract

Questions 368-369

A young man, recently returned to the United States from Thailand, has severe liver disease. Symptoms include jaundice, anemia, and weakness.

368. The etiologic agent, shown in the photomicrographs on the facing page, is

(A) *Plasmodium falciparum*
(B) *Clonorchis sinensis*
(C) *Diphyllobothrium latum*
(D) *Taenia solium*
(E) *Taenia saginata*

369. An intermediate form of the organism shown on the facing page lives in

(A) mosquitoes
(B) pigs
(C) snails
(D) cows
(E) ticks

370. Recommendations for the control of human hookworm in endemic areas include the construction of sanitary facilities and the

(A) thorough washing of fresh fruit and vegetables
(B) thorough cooking of all meats
(C) reduction of the wild dog population
(D) use of insecticides to control flies
(E) wearing of footwear

Photomicrographs accompany Questions 368-369

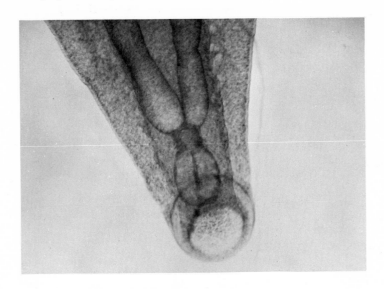

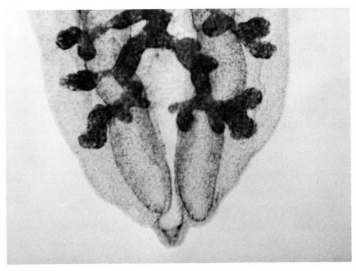

371. In the typical life cycle of a trematode (e.g., *Schistosoma*), which of the following developmental forms enters the intermediate snail host?

(A) Cercaria
(B) Metacercaria
(C) Schizont
(D) Redia
(E) Miracidium

372. An experimental antibiotic specific for protozoal infection could be expected to have the LEAST significant effect on which of the following parasites?

(A) *Entamoeba coli*
(B) *Echinococcus granulosus*
(C) *Pneumocystis carinii*
(D) *Plasmodium vivax*
(E) *Giardia lamblia*

373. One million persons in the United States have roundworm infection. Which of the following parasites is a roundworm that hatches in the upper small intestine and releases rhabditiform larvae that penetrate the intestinal wall?

(A) *Hymenolepis nana*
(B) *Diphyllobothrium latum*
(C) *Schistosoma mansoni*
(D) *Fasciola hepatica*
(E) *Ascaris lumbricoides*

374. Which of the following parasitic infections is most common in the continental United States?

(A) Ascariasis
(B) Enterobiasis
(C) Trichinosis
(D) Schistosomiasis
(E) Trypanosomiasis

375. Organisms of the type depicted below are seen in diarrheic feces. This finding is most compatible with a diagnosis of

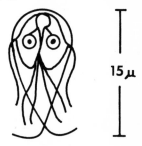

15 μ

(A) bilharziasis
(B) ascariasis
(C) enterobiasis
(D) giardiasis
(E) shigellosis

376. Which of the following is the drug of choice for the treatment of individuals who have ascariasis?

(A) Niridazole (Ambilhar)
(B) Piperazine citrate (Antepar)
(C) Pyrvinium pamoate (Povan)
(D) Quinacrine hydrochloride (Atabrine)
(E) Thiabendazole (Mintezol)

377. The drug of choice for treating persons infested with *Taenia saginata* (beef tapeworm) is

(A) emetine hydrochloride
(B) niclosamide (Yomesan)
(C) piperazine citrate (Antepar)
(D) quinacrine hydrochloride (Atabrine)
(E) thiabendazole (Mintezol)

378. Analysis of a patient's stool reveals small structures resembling rice grains; microscopic examination shows these to be "proglottids." The most likely organism in this patient's stool is

(A) *Enterobius vermicularis*
(B) *Ascaris lumbricoides*
(C) *Necator americanus*
(D) *Taenia saginata*
(E) *Trichuris trichiura*

379. In the diagnosis of kala-azar, which of the following tests is most likely to reveal active disease?

(A) Liver biopsy
(B) Blood culture
(C) Splenic aspiration
(D) Stool examination
(E) Sigmoidoscopy

380. Which of the following parasites, transmitted in cat feces, poses the greatest risk to the unborn infants of pregnant women?

(A) *Clonorchis*
(B) *Enterobius*
(C) *Wuchereria*
(D) *Toxocara*
(E) *Toxoplasma*

381. Microscopic examination of blood and lymph would be LEAST likely to reveal

(A) *Brugia malayi*
(B) *Loa loa*
(C) *Onchocerca volvulus*
(D) *Schistosoma haematobium*
(E) *Wuchereria bancrofti*

382. A woman recently returned from Africa complains of having paroxysmal attacks of chills, fever, and sweating; these attacks last a day or two at a time and recur every 36 to 48 hours. Examination of a stained blood specimen reveals ring-like and crescent-like forms within red blood cells. The infecting organism most likely is

(A) *Plasmodium falciparum*
(B) *Trypanosoma gambiense*
(C) *Wuchereria bancrofti*
(D) *Plasmodium vivax*
(E) *Schistosoma mansoni*

383. Which of the following parasites may be ingested in uncooked fish?

(A) *Hymenolepis diminuta*
(B) *Taenia saginata*
(C) *Diphyllobothrium latum*
(D) *Strongyloides stercoralis*
(E) *Schistosoma japonicum*

384. A woman who recently traveled through Central Africa now complains of severe chills and fever, abdominal tenderness, and darkening urine. Her febrile periods last for 28 hours and recur regularly. Which of the blood smears drawn below would be most likely to be associated with the symptomatology described?

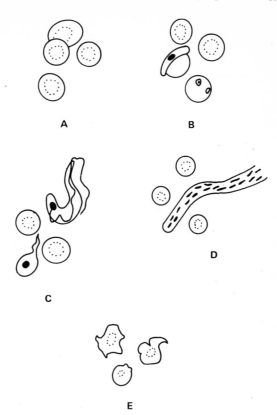

A

B

C

D

E

(A) A
(B) B
(C) C
(D) D
(E) E

DIRECTIONS: Each question below contains four suggested answers of which **one** or **more** is correct. Choose the answer:

A	if	**1, 2, and 3**	are correct
B	if	**1 and 3**	are correct
C	if	**2 and 4**	are correct
D	if	**4**	is correct
E	if	**1, 2, 3, and 4**	are correct

385. The clinical signs and symptoms of trichinosis include

(1) eosinophilia
(2) periorbital edema
(3) muscle pain
(4) diarrhea

386. The finding of greater than 3 to 5 percent eosinophils in the peripheral blood is

(1) a consistent sign of parasitic infection
(2) not noted in other forms of infection
(3) an invaluable diagnostic indicator of parasitic infection
(4) more marked in recent than in chronic parasitic infection

387. The bedbugs *Cimex lectularius* and *C. hemipterus* can be described by which of the following statements?

(1) They are bloodsucking parasites of humans
(2) They can be vectors for Chagas' disease
(3) They can be destroyed by the pesticide DDT
(4) They produce a diffuse erythematous rash

388. Which of the following statements about human lice are true?

(1) They are wingless
(2) They cause pruritic skin lesions
(3) They transmit epidemic typhus, relapsing fever, and trench fever
(4) *Pediculus humanus* and *Phthirus pubis* are two species

389. A survey of 100 healthy adults reveals that 80 percent have antibodies to *Toxoplasma*. Which of the following statements would help to explain this finding?

(1) The potential for *Toxoplasma* infection is widespread
(2) Toxoplasmosis can be an extremely mild, self-limiting disease in adults
(3) A large percentage of healthy adults have been exposed to *Toxoplasma*
(4) The test for *Toxoplasma* antibodies is highly nonspecific

390. Scabies is caused by a small mite that burrows into the skin. The disease is

(1) caused by a species of *Sarcoptes*
(2) often complicated by secondary bacterial infection
(3) best diagnosed by morphologic identification of the mite
(4) effectively treated with gamma benzene hexachloride

391. The parasite causing hydatid disease is the cysticercus form of *Echinococcus granulosus*. Which of the following statements about this parasite are true?

(1) It frequently affects the liver
(2) It is part of the normal intestinal flora
(3) Its adult form commonly involves the dog as an intermediate host
(4) It often can be recovered from human feces

392. Which of the following statements about parasitic trematode infections are true?

(1) Affected persons are treated effectively with thiabendazole, a broad-spectrum antihelminthic agent
(2) One or two intermediate hosts may be involved
(3) Humans typically are affected by eating undercooked, infected beef
(4) These infections are rarely contracted in the continental United States

393. Which of the following organisms can infect humans who eat improperly prepared meat?

(1) *Taenia solium*
(2) *Trichinella spiralis*
(3) *Taenia saginata*
(4) *Diphyllobothrium latum*

394. Helminths that principally infest the liver include

(1) *Fasciola*
(2) *Ascaris*
(3) *Clonorchis*
(4) *Paragonimus*

395. Transfusion malaria can be described by which of the following statements?

(1) It is a benign form of malaria, with no associated mortality
(2) It is due primarily to *Plasmodium falciparum* in the United States
(3) It is characterized by an exoerythrocytic liver cycle
(4) It usually responds readily to treatment

396. The photomicrograph below shows fine fibrils (labeled "F") in an ameba. These structures are

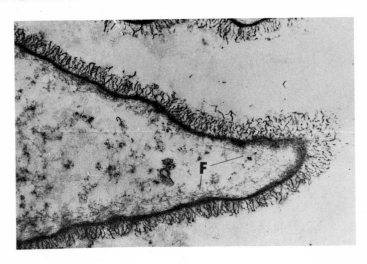

(1) termed "amebic microtubules"
(2) analogous to actin in the cells of higher forms of life
(3) primarily of glycoprotein composition
(4) involved in cell motility

---

397. A̶d̶u̶l̶t̶ tapeworms are injurious *flatworms - cestodes* to the human host because they

(1) may deprive the host of vitamins
(2) obtain protein from the host's intestinal mucosa
(3) can predispose to bacterial invasion at the site of scolex attachment
(4) produce an enterotoxin causing malabsorption

398. Which of the following statements about *Giardia lamblia* are true?

(1) It is a flagellated protozoon
(2) It is frequently found in water supplies contaminated with sewage
(3) It has been implicated in some cases of cholecystitis
(4) It is primarily a colonic parasite

SUMMARY OF DIRECTIONS

A	B	C	D	E
1, 2, 3 only	1, 3 only	2, 4 only	4 only	All are correct

399. The three species of the proto-zoon *Trichomonas* that can infect humans can be described by which of the following statements?

(1) *T. hominis* is the most pathogenic for humans
(2) *T. tenax* infects the distal small intestine
(3) *T. vaginalis* usually causes erosion of the uterine mucosa
(4) *T. vaginalis* is transmitted sexually

400. In the asexual cycle of *Plasmodium*, the length of time from the merozoite stage of one cycle to the merozoite stage of the next is 50 hours or less for

(1) *P. vivax*
(2) *P. falciparum*
(3) *P. ovale*
(4) *P. malariae*

401. Pinworm infections may spread rapidly in a family. These infections are

(1) characteristic exclusively of human hosts
(2) widely distributed throughout the world
(3) common in the continental United States
(4) most likely to affect young children

402. Schistosomiasis is a disease characterized by granulomatous re-actions to the ova or to products of the parasite at the place of oviposition. Other clinical manifestations can include

(1) calcifications in the bladder wall
(2) pulmonary arterial hypertension
(3) splenomegaly
(4) esophageal varices

403. *Entamoeba histolytica* is a human intestinal parasite that

(1) usually causes asymptomatic infection
(2) is 15 to 30 μm in size
(3) pathognomonically contains red blood cells in the cytoplasm
(4) appears as trophozoites in the stool of asymptomatic carriers

404. Which of the following state-ments about plasmodia are true?

(1) The schizogonic cycle of plasmodia occurs in the *Anopheles* mosquito
(2) The first stage of human plasmodial infection is exoerythrocytic
(3) Coarse stippling of erythrocytes is characteristic of *Plasmodium vivax* infection
(4) *P. falciparum* invades red blood cells of any age

405. *Trypanosoma cruzi* is associated with

(1) interstitial myocarditis
(2) unilateral swelling of the eyelids
(3) less parasitemia than occurs with *T. gambiense*
(4) tsetse-fly transmission

406. Amebas found in the intestines of humans are

(1) often nonpathogenic
(2) pathogenic when cysts become toxigenic
(3) a cause of appendicitis
(4) transmitted usually as trophozoites

DIRECTIONS: The group of questions below consists of lettered choices followed by several numbered items. For each numbered item select the **one** lettered choice with which it is **most** closely associated. Each lettered choice may be used once, more than once, or not at all.

Questions 407-411

Match the following case histories with the most likely diagnosis.

(A) Trichinosis
(B) Schistosomiasis
(C) Toxoplasmosis
(D) Visceral larva migrans
(E) Giardiasis

407. A butcher, who is fond of eating raw hamburger, develops chorioretinitis; a Sabin-Feldman dye test is positive

408. A fur trapper complains of sore muscles, has swollen eyes, and reports eating bear meat on a regular basis

409. A correspondent for *The New York Times* has diarrhea for two weeks following a trip to Leningrad

410. A retired Air Force colonel has had abdominal pain for two years; he makes yearly freshwater-fishing trips to Puerto Rico and often wades with bare feet into streams

411. A teenager who works in a dog kennel after school has had a skin rash, eosinophilia, and an enlarged liver and spleen for two years

Parasitology

Answers

357. The answer is D. *(Brown, ed 4. pp 97-98. Jawetz, ed 13. pp 504-507.)* The large reservoir of malarial disease in the United States was eliminated by vigorous treatment of all human cases as well as by effective mosquito control. The discovery that quinine and its derivatives are effective in treating malaria was a major medical breakthrough; today, the treatment of choice is chloroquine. Malaria is spread by *Anopheles*, not *Culex*, mosquitoes.

358. The answer is B. *(Jawetz, ed 13. p 494.)* *Giardia lamblia* is a protozoon that can infest the human duodenum and jejunum. Symptoms of giardiasis include weakness, malaise, abdominal distension and cramps, diarrhea, and steatorrhea. In symptomatic patients, large numbers of *Giardia* cysts may be found in formed or liquid stool; these elliptical cysts are 10 to 14 μm in length and contain 2 to 4 nuclei. The heart-shaped trophozoite can be isolated from liquid stool only.

359. The answer is E. *(Brown, ed 4. p 1.)* Facultative parasites are capable of sustaining growth on their own in an extracellular environment or within a cell. For obligate parasites to grow and reproduce, they must reside on or within another living organism. A pseudoparasite is not a true parasite but an artifact mistaken for a true parasite.

360. The answer is A. *(Brown, ed 4. p 195.)* The adult form of *Echinococcus granulosus* lives in the intestine of dogs, wolves, foxes, and other carnivores. Humans become infected by ingesting eggs of the parasite. The eggs hatch in the intestine and migrate to the lungs and liver, where the characteristic hydatid cysts form.

361. The answer is A. *(Brown, ed 4. p 111.)* A specific diagnosis of infection with *Trichinella spiralis* can be made by demonstration of larvae in a muscle-biopsy specimen. Examination of the feces of an affected individual for adult worms and of the blood, cerebrospinal fluid, and other body fluids for larvae usually is negative. Intradermal and complement-fixation tests are not completely accurate. Eosinophilia, myositis, and periorbital edema are important diagnostic clues to the presence of trichinosis.

362. The answer is D. *(Brown, ed 4. pp 146-147.)* Filariasis, also known as elephantiasis and wuchereriasis, is caused by the extensive growth and proliferation of *Wuchereria bancrofti*. The parasite is transmitted through the bite of mosquitoes, and humans are the only definitive hosts. Microfilariae are common in blood and can be observed in blood smears from infected individuals.

363. The answer is C. *(Brown, ed 4. p 119.)* *Strongyloides* is a common intestinal parasite in tropical areas. The rhabditiform larvae of *S. stercoralis* may develop into infective filariform larvae in transit down the bowel and produce reinfection by invading the mucosa of the lower portion of the colon. Because of autoinfection, strongyloidiasis, if untreated, may persist for years.

364. The answer is E. *(Brown, ed 4. p 132.)* The pinworm, *Enterobius vermicularis*, is a parasite of the cecum and intestine. At night, female worms migrate to deposit eggs in perianal and perineal regions. Eggs can be recovered easily by perianal swabbing with cellophane tape; the eggs adherent to the tape can be identified microscopically. Swabbing for three consecutive days reveals the presence of eggs in 90 percent of cases. Recovery of pinworm eggs from fecal specimens is often unrewarding.

365. The answer is D. *(Brown, ed 4. pp 54-58.)* *Trypanosoma cruzi*, the cause of Chagas' disease, can be transmitted by 28 species of reduviid bug (the principal vector is *Triatoma*). The bug often defecates while biting; the fecal material, which contains the infective metacyclic stage of the trypanosome, can contaminate the bite wound. Blood transfusions can also be a source of infection. At present, there is no totally effective treatment available for individuals suffering from Chagas' disease.

366. The answer is C. *(Brown, ed 4. pp 162-163.)* In an unnatural host (e.g., humans) the larvae of the dog ascarid (*Toxocara canis*) and the cat ascarid (*T. cati*) are unable to complete their normal development. Instead, they migrate extensively through extraintestinal viscera until they are stopped by a host cell response, which isolates them in granulomas. Lesions containing *Toxocara* larva have been found in the eye, brain, liver, kidney, lungs, and lymph nodes.

367. The answer is B. *(Faust, ed 8. pp 529-538.)* Both beef tapeworm (*Taenia saginata*) and pork tapeworm (*T. solium*) can, in the adult form, cause disturbances of intestinal function. Intestinal disorder is due not only to direct irritation but also to the action of metabolic toxic wastes. In addition, *T. saginata*, because of its large size, may produce acute intestinal blockage. Unlike *T. saginata*, *T. solium* produces cysticercosis which results in serious lesions in humans (in *T. saginata*, the cysticercus—encysted larvae—stage develops only in cattle).

368. The answer is B. *(Brown, ed 4. pp 226-230.)* The Chinese liver fluke, *Clonorchis sinensis*, is a parasite of humans and is found in Japan, China, South Korea, Taiwan, and Indochina. Humans usually are infected by eating uncooked fish. The worms invade bile ducts and produce destruction of liver parenchyma. Anemia, jaundice, weakness, weight loss, and tachycardia may follow. Treatment is likely to be ineffectual in heavy infections, but chloroquine can destroy some of the worms.

369. The answer is C. *(Brown, ed 4. pp 226-227.)* The life cycle of *Clonorchis sinensis* is similar to that of other trematodes. A mollusk is characteristically the first intermediate host for trematodes. For *C. sinensis*, snails perform this role.

370. The answer is E. *(Brown, ed 4. pp 125-126.)* *Necator americanus* and *Ancylostoma duodenale* are responsible for most human hookworm infections. Filariform larvae of these organisms gain access to humans by penetrating skin, usually of the interdigital spaces between the toes of individuals who are barefoot. Warm climate, accessibility of fecal matter, and damp, loosely packed soil are ideal conditions for the growth and spread of hookworms.

371. The answer is E. *(Brown, ed 4. p 214.)* In the typical life cycle of trematodes, eggs are discharged from the intestinal or genitourinary tract of a definitive host. The eggs hatch in fresh water, releasing the larval miracidia, which enter the snails that serve as intermediate hosts. By metamorphosis, miracidia become rediae, which in turn develop into cercariae. The cercariae are released from the intermediate host and re-enter the water. To cause human infection, encysted metacercariae must be ingested; on the other hand, cercariae can penetrate skin. The schizont is an asexual form of malarial protozoa and is not a developmental form of trematodes.

372. The answer is B. *(Brown, ed 4. pp 6-10.)* The protozoa that infect humans include amebas (e.g., *Entamoeba coli*), flagellates (e.g., *Giardia lamblia*), trypanosomes (e.g., *Leishmania donovani*), sporozoa (e.g., *Plasmodium vivax*), and other protozoal forms of uncertain nature (e.g., *Pneumocystis carinii*). *Echinococcus granulosus* is a tapeworm that causes hydatid disease in humans. Thus, an antiprotozoal agent would not be expected to be effective against echinococcal infection.

373. The answer is E. *(Brown, ed 4. pp 4-8, 209, 231.)* Helminths are subdivided into three phyla: the Annelida, or segmented worms; the Nemathelminthes, or roundworms; and the Platyhelminthes, or flatworms. *Ascaris lumbricoides* is a roundworm that hatches in the upper small intestine of infected humans; rhabditiform larvae are released and penetrate the intestinal wall. The other four parasites listed are flatworms: *Diphyllobothrium latum* and *Hymenolepis nana* are cestodes (tapeworms), and *Fasciola hepatica* and *Schistosoma mansoni* are trematodes (flukes).

374. The answer is B. *(Brown, ed 4. pp 8-9, 54, 94, 108, 131.)* Pinworm, an infection with *Enterobius vermicularis*, is a widespread and exceedingly common disease (surveys in Washington, D.C., show a 12 to 41 percent infection rate). Trichinosis affects approximately 2 percent of the American population, compared to 15 to 25 percent 35 years ago. Schistosomiasis and trypanosomiasis are, for the most part, limited to Africa, South America, and Asia. Ascariasis is common in many countries worldwide.

375. The answer is D. *(Brown, ed 4. pp 40-41.)* The organism sketched in the question is too small for a worm and too large for a bacterium. It is the trophozoite form of *Giardia lamblia*. Giardiasis can cause acute diarrhea, abdominal pain, and weight loss. It is spread through contaminated food and water.

376. The answer is B. *(Brown, ed 4. p 137.)* Piperazine citrate (Antepar) and other piperazine salts have become the drugs of choice in the treatment of disease caused by roundworms, including *Ascaris*. The drugs are safe and 95 percent effective when administered during two consecutive days. They act by relaxing worm musculature, causing the worm to be eliminated by normal intestinal peristaltic action.

377. The answer is B. *(Brown, ed 4. pp 182, 191.)* Niclosamide (Yomesan) is the drug of choice in the treatment of infection with several tapeworm species, including *Taenia saginata* and *Diphyllobothrium latum*. It is also effective against

Taenia solium, but only when given with a saline cathartic and an antiemetic agent to avoid cysticercosis. Although quinacrine hydrochloride (Atabrine) is effective for treating tapeworm infection, it frequently causes vomiting.

378. The answer is D. *(Brown, ed 4. pp 105-141, 178-194.)* *Enterobius* (pinworm), *Ascaris* (roundworm), *Necator* (hookworm), and *Trichuris* (whipworm) are roundworms, or nematodes. *Taenia saginata* (tapeworm), a segmented flatworm, affects the small intestine of humans. Tapeworm segments, called "proglottids," appear in the stool of infected individuals.

379. The answer is C. *(Brown, ed 4. pp 61-62. Jawetz, ed 13. p 497.)* *Leishmania donovani*, the causative agent of kala-azar, multiplies in reticuloendothelial cells, especially in macrophages of the spleen, lymph nodes, and bone marrow. Blood usually is examined first, when the disease is suspected, but skillful splenic puncture has a much higher percentage of positive findings and thus is the method of choice. Biopsies of the liver, lymph nodes, and bone marrow are also useful diagnostic procedures.

380. The answer is E. *(Brown, ed 4. p 68. Jawetz, ed 13. pp 508-509.)* Toxoplasmosis is a disease caused by *Toxoplasma gondii*, a protozoon distributed worldwide. Although infection with this organism is common (up to 80 percent in some populations), actual symptomatic disease is rare. *T. gondii* is transmitted by contact with raw meat or cat feces infected with oocysts. Congenital toxoplasmosis acquired during the first trimester of pregnancy is associated with profound anomalies of the fetal central nervous system.

381. The answer is D. *(Brown, ed 4. p 142.)* Microfilariae are parasitic nematodes of the blood and lymphatics. *Wuchereria bancrofti*, the best-known member of this group, frequently causes inflammation of the lymphatics; obstructive edema and elephantiasis can ensue. Other microfilariae include *Brugia malayi*, *Loa loa*, and *Onchocerca volvulus*. *Schistosoma haematobium* is a trematode, not a microfilaria, and is found usually in the urine.

382. The answer is A. *(Brown, ed 4. pp 82-83, 88-89.)* The febrile paroxysms of *Plasmodium malariae* malaria occur at 72-hour intervals; those of *P. falciparum* and *P. vivax* malaria occur every 48 hours. The paroxysms usually last 8 to 12 hours with *P. vivax* malaria but can last 16 to 36 hours with *P. falciparum* disease. In *P. vivax*, *P. ovale*, and *P. malariae* infection, all stages of development of the organisms can be seen in the peripheral blood; in malignant tertian (*P. falciparum*) infections, only early ring stages and gametocytes are usually found.

383. The answer is C. *(Brown, ed 4. pp 179-181.)* *Diphyllobothrium latum* commonly is transmitted to humans by ingestion of infected fresh-water fish. The parasite can be found on nearly every continent. *Taenia saginata* is transmitted in uncooked beef. *Hymenolepis diminuta* is spread by infected insects. Free-living forms of *Strongyloides* and *Schistosoma* directly penetrate the skin of their hosts.

384. The answer is B. *(Brown, ed 4. pp 79, 82.)* The case history presented in the question is characteristic of infection with *Plasmodium falciparum*, the causative agent of malignant tertian malaria. The long duration of the febrile stage rules out other forms of malaria. The presence of ring-form young trophozoites and crescent-form mature gametocytes—as represented in the illustration below—and the absence of schizonts are diagnostic of *P. falciparum* malaria.

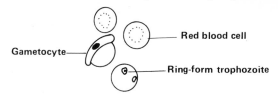

385. The answer is E (all). *(Brown, ed 4. pp 109-110.)* The clinical course of *Trichinella spiralis* disease (trichinosis) progresses through three phases, which correspond to stages in the life cycle of the infecting organism. During the first phase, intestinal invasion by adult worms clinically manifests as diarrhea and abdominal pain. Migration of the larvae then occurs, accompanied by muscle pain, eosinophilia, and edema. Finally, with encystment of the parasites, weakness and fatigue ensue.

386. The answer is D (4). *(Brown, ed 4. p 318.)* Too much emphasis often is placed on eosinophilia as a definite sign of parasitic disease. Eosinophilia is found in a variety of diseases, including asthma, gonorrhea, varicella-zoster, periarteritis nodosa, carcinoma of the rectum, and many others. Eosinophilia is not a consistent sign of parasitic disease; the degree of eosinophilia varies with both host response and duration of infection (it is more marked in recent than in chronic parasitic infection).

387. The answer is B (1, 3). *(Brown, ed 4. pp 268-270.)* Bedbugs *(Cimex)* are bloodsucking parasites of humans. At night they leave their hiding places to feed on the blood of humans and small mammals. Their bite produces pruritic red wheals, which characteristically are paired. Bedbugs have not been proven to be vectors of human disease. The pesticide DDT applied to furniture and mattresses can control these parasites; because of its environmental impact, however, this agent may be quite hard to procure.

388. The answer is E (all). *(Brown, ed 4. pp 260-262.)* Pediculus humanus (head or body louse) and *Phthirus pubis* (crab louse) are parasites exclusively affecting humans. Lice are important not only for the itching and discomfort they cause but also for the diseases they transmit. These disorders include epidemic typhus, relapsing fever, and trench fever.

389. The answer is A (1, 2, 3). *(Brown, ed 4. pp 68-69.)* Serologic tests, such as the Sabin-Feldman dye test and indirect immunofluorescence, have shown that a high percentage of the world's population has been infected with *Toxoplasma gondii*. In adults, clinical toxoplasmosis usually presents as a benign syndrome resembling infectious mononucleosis. However, fetal infections are often severe and associated with hydrocephalus, chorioretinitis, convulsions, and death.

390. The answer is E (all). *(Brown, ed 4. pp 307-309.)* Sarcoptes scabiei is a small mite that burrows into human skin. Itching is significant, and a vesicular eruption, which often becomes secondarily infected with bacteria, develops. Diagnosis is made by microscopic detection of the mites. Gamma benzene hexachloride (Kwell), a topical insecticide, is an effective treatment for scabies.

391. The answer is B (1, 3). *(Brown, ed 4. pp 195-202.)* Hydatidosis is caused by the cysticercus form of the flatworm *Echinococcus granulosus*. Adult worms are harbored in the small intestine of dogs, coyotes, and other intermediates, and eggs are transmitted in the feces of these hosts. Echinococcosis, a rare infection in the United States, typically involves the liver; brain, lungs, and bones are involved less commonly.

392. The answer is C (2, 4). *(Brown, ed 4. pp 209-254.)* Trematode infections are rarely contracted in the United States (fewer than ten cases per year), although many "imported" cases occur yearly. The intermediate host of the parasitic trematodes is, typically, crustacean—either a single crustacean, as in the case of schistosomes, or two crustaceans (or a crustacean and aquatic vegetation), as in the case of the hermaphroditic flukes. The treatment of individuals infected with trematodes is difficult and nonspecific; thiabendazole is an antinematode drug and thus would be an ineffective remedy.

393. The answer is A (1, 2, 3). *(Brown, ed 4. pp 105-112, 187-193.)* The pork and beef tapeworms *(Taenia solium* and *T. saginata,* respectively)as well as *Trichinella spiralis* all spend a portion of their life cycle encysted in the tissues of secondary hosts that are common human foods. These organisms remain infectious if they are not killed by adequate freezing or heating of the infected meat. *Diphyllobothrium latum* is a type of tapeworm that can be ingested with uncooked freshwater fish.

394. The answer is B (1, 3). *(Brown, ed 4. pp 133-137, 162-164.)* The liver flukes, so named because of their predisposition for biliary infestation, include *Echinococcus, Clonorchis, Fasciola,* and *Opisthorchis.* Paragonimiasis is chiefly a pulmonary fluke infection; fibrosis develops, and hemoptysis is a cardinal sign. Ascariasis is mainly an intestinal disease in humans, most of whom are only minimally symptomatic when infected; migrating roundworms, however, can infect the liver and lungs, and on emergence from the upper airway, cause aspiration.

395. The answer is D (4). *(Brown, ed 4. p 86.)* Malaria can be transmitted both by transfusion from an infected donor and by inoculation with an infected syringe. In the United States, *Plasmodium malariae* is the most common cause of transfusion malaria. Transfusion malaria is not associated with exoerythrocytic liver cycles. Although the organisms usually are easy to eradicate, the illness can be fatal, depending on the inoculum.

396. The answer is C (2, 4). *(Ishikawa, J Cell Biol 43 [1969] :312. Jawetz, ed 13. pp 500-501.)* The movement of amebic trophozoites is usually unidirectional and controlled in part by chemotactic factors in the immediate environment. Amebas are motile by virtue of pseudopods, which are cytoplasmic extensions that alternately project and contract. The fibrils shown in the amebic pseudopod pictured in the question are involved in cellular motility and are similar to muscle actin. Amebic motility is retarded as environmental temperature falls below 37°C.

397. The answer is A (1, 2, 3). *(Brown, ed 4. pp 176-177.)* Disease produced by adult cestodes is not usually clinically significant. Minor clinical disturbances, such as weakness, fatigue, and irritability, can result from vitamin deficiency, local breakdown of the intestinal mucosa, bacterial superinfection, and other factors. Infrequently, tapeworm disease can lead to intestinal obstruction or perforation, when the worms are present in great numbers.

398. The answer is A (1, 2, 3). *(Jawetz, ed 13. p 494.)* *Giardia lamblia* is a flagellated protozoon that infects the duodenum and jejunum of humans and can cause flagellate diarrhea or giardiasis. Mild cholecystitis also may occur, if the bile ducts and gallbladder are invaded. *G. lamblia* is transmitted in fecally contaminated water; epidemic outbreaks of giardiasis have occurred. Treatment with quinacrine hydrochloride (Atabrine) or metronidazole (Flagyl) is curative in the majority of clinically significant cases.

399. The answer is D (4). *(Jawetz, ed 13. p 495.)* Three species of *Trichomonas* are known to infect humans. *T. tenax* is found in the mouth, *T. hominis* in the intestine, and *T. vaginalis* in the genitourinary tract. *T. hominis* and *T. tenax* are considered nonpathogenic, whereas *T. vaginalis* is responsible for vaginal (but not uterine) infections in women and prostatic and urethral infections in men. Transmission of *T. vaginalis* occurs during coitus. Trichomoniasis in both women and men can be cured by the administration of metronidazole (Flagyl).

400. The answer is A (1, 2, 3). *(Brown, ed 4. pp 75-98. Jawetz, ed 13. pp 504-506.)* The length of the asexual cycle of *Plasmodium vivax, P. ovale,* and *P. falciparum* is 48 hours or less. In the case of *P. malariae*, 72 hours are required for the reproduction of merozoites. The periodicity of chills, fever, nausea, and vomiting in malarial infections corresponds to the end of the schizogonic cycle, when merozoites of the mature schizonts rupture into the circulation; thus, these symptoms appear every other day with *P. vivax, P. ovale,* and *P. falciparum* infection but every 72 hours with *P. malariae* disease.

401. The answer is E (all). *(Brown, ed 4. pp 129-133.)* Pinworm infection is innocuous and usually self-limited. This common disease most often affects children, who may spread the infection to other household members. Pyrvinium pamoate (Povan) or piperazine should be given to every member of the family of an affected child.

402. The answer is E (all). *(Brown, ed 4. pp 239-252.)* Although the chronic stage of proliferation within tissues is distinctive in the different forms of schistosomiasis, a granulomatous reaction to the eggs and chemical products of the schistosome occurs in all forms of the disease. *Schistosoma haematobium* commonly involves the distal bowel and the bladder, as well as the prostate gland and seminal vesicles. Bladder calcification and cancer may ensue. *S. mansoni* affects the large bowel and the liver; presinusoidal portal hypertension, splenomegaly, and esophageal varices may be complications. Pulmonary hypertension, often fatal, may be seen with *S. mansoni* and *S. japonicum* disease. Eggs may be found in an unstained specimen of rectal mucosa or in stool. Urine microscopy and liver biopsy, where warranted, often prove positive. The treatment of choice for schistosomiasis is preventive: i.e., elimination of the parasite in snails, before human infection occurs.

403. The answer is A (1, 2, 3). *(Jawetz, ed 13. pp 500-502.)* *Entamoeba histolytica* is an intestinal parasite of humans; its size ranges from 15 to 30 μm and its cytoplasm characteristically contains red blood cells. It usually causes asymptomatic infection but may cause amebic dysentery. The cysts of *E. histolytica* may be seen even in the stool of asymptomatic individuals. Dysentery and symptoms of amebiasis occur only after trophozoites invade the intestinal mucosa; at this point, trophozoites also appear in the stool.

404. The answer is D (4). *(Jawetz, ed 13. pp 504-507.)* Malaria involves a sporogonic sexual cycle, which takes place in the female *Anopheles* mosquito, and a schizogonic asexual cycle, which takes place in humans. The first phase of human infection is pre-erythrocytic, occurring in the liver parenchyma and yielding merozoites, which enter the bloodstream and infest erythrocytes. Certain species of plasmodia (*P. vivax*, for example) typically infect red blood cells of a particular age; however, *P. falciparum* invades red blood cells of all ages and, thus, is the most severe type of malaria. Cells parasitized with *P. falciparum* have coarse stippling; fine stippling of parasitized cells is associated with *P. vivax* and *P. ovale*.

405. The answer is A (1, 2, 3). *(Jawetz, ed 13. pp 498-500.)* American trypanosomiasis (Chagas' disease) is produced by *Trypanosoma cruzi*, which is transmitted to humans in the feces of the reduviid bug percutaneously or by conjunctival infection. Inoculation by the latter route may result in Romaña's sign: unilateral eyelid swelling. Hepatosplenomegaly results from infection involving the reticuloendothelial system; interstitial myocarditis results from involvement of the heart and may cause congestive heart failure. African trypanosomiasis, or sleeping sickness, is caused by *T. gambiense* or *T. rhodesiense*; infection with these parasites, which are transmitted by the tsetse fly, results in greater parasitemia than is associated with *T. cruzi*.

406. The answer is B (1, 3). *(Jawetz, ed 13. pp 500-504.)* Amebic infestation of the human intestine is often asymptomatic. Symptomatic amebiasis and dysentery result when ameboid trophozoites invade the intestinal wall, producing ulceration and diarrhea; extraintestinal involvement, e.g., hepatic or brain abscess, occasionally develops. Amebic disease usually is transmitted by the cyst form of the organism.

407-411. The answers are: 407-C, 408-A, 409-E, 410-B, 411-D. *(Brown, ed 4. pp 41-42, 162-164. Cahill, pp 56-62, 90-91, 94-102.)* All of the diseases listed in the question have significant epidemiologic and clinical features. Toxoplasmosis, for example, is generally a mild, self-limiting disease; however, severe fetal disease is possible if pregnant women ingest *Toxoplasma* oocysts. Consumption of uncooked meat may result in either an acute toxoplasmosis or a chronic toxoplasmosis that is associated with serious eye disease. Most adults have antibody titers to *Toxoplasma* and thus would have a positive Sabin-Feldman dye test.

Trichinosis most often is caused by ingestion of contaminated pork products. However, eating undercooked bear, walrus, raccoon, or possum meat also may cause this disease. Symptoms of trichinosis include muscle soreness and swollen eyes.

Schistosomiasis is a worldwide public-health problem. Control of this disease entails the control of the snail intermediate host and removal of stream-side vegetation. Abdominal pain is a symptom of schistosomiasis.

Although giardiasis has been classically associated with travel in the Soviet Union, especially Leningrad, many cases of giardiasis due to contaminated water have been reported in the United States as well. Diagnosis is made by detecting cysts in the stool. In some cases, diagnosis may be very difficult because of the relatively small numbers of cysts present.

Visceral larva migrans is an occupational disease of individuals who are in close contact with dogs and cats. The disease is caused by the nematodes *Toxocara canis* (dogs) and *T. cati* (cats) and has been recognized in young children who have close contact with pets or who eat dirt. Symptoms include skin rash, eosinophilia, and hepatosplenomegaly.

Immunology

DIRECTIONS: Each question below contains five suggested answers. Choose the **one best** response to each question.

412. Which of the following substances is generated chiefly by B lymphocytes (B cells)?

(A) Antistreptolysin O
(B) Lymphotoxin
(C) Blastogenic factor
(D) Migration-inhibition factor
(E) Transfer factor

413. The class-specific antigenic determinant of immunoglobulins is associated with the

(A) J chain
(B) T chain
(C) light chain
(D) heavy chain
(E) secretory component

414. Bence Jones proteins, which are often found in the urine of individuals who have multiple myeloma, are best described as

(A) mu chains
(B) gamma chains
(C) kappa and lambda chains
(D) albumin derivatives
(E) fibrin split products

415. A young girl has had repeated infections with *Candida albicans* and respiratory viruses since the time she was three months old. As part of the clinical evaluation of her immune status, her responses to routine immunization procedures should be tested. In this evaluation, the use of which of the following vaccines is contraindicated?

(A) Diphtheria toxoid
(B) *Bordetella pertussis* vaccine
(C) Tetanus toxoid
(D) BCG
(E) Inactivated polio

416. Which of the following diseases is associated with latent measles-virus infection and, presumably, a defect in cellular immunity?

(A) Infectious mononucleosis
(B) Creutzfeldt-Jakob disease
(C) Kuru
(D) Subacute sclerosing panencephalitis
(E) Burkitt's lymphoma

417. Antibodies that neutralize the infectivity of influenza virus are formed against which of the following virus antigens?

(A) Nucleocapsid
(B) Nucleoprotein
(C) Neuraminidase
(D) M protein on the viral envelope
(E) Hemagglutinin on the viral surface

418. Which of the following tests is the most sensitive for measuring anti-viral antibody?

(A) Neutralization
(B) Double diffusion in agar gel
(C) Complement fixation
(D) Radioimmunoassay
(E) Immunoelectrophoresis

419. The class of immunoglobulin (Ig) important in protecting the mucosal surfaces of the respiratory, intestinal, and genitourinary tracts from pathogenic organisms is

(A) IgA
(B) IgD
(C) IgE
(D) IgG
(E) IgM

420. Children who have received in-activated measles vaccine may have swelling, tenderness, and erythema at the site of a subsequent live measles vaccination. If this reaction is at its maximum approximately four hours after the live measles immunization, it is most likely to be a type of

(A) Prausnitz-Küstner reaction
(B) Arthus reaction
(C) Shwartzman reaction
(D) Schick reaction
(E) Coombs reaction

421. Complement is a complex series of interacting proteins able to amplify the actions of certain classes of anti-bodies. Which of the functional units of complement must interact to cause swift lysis of antibody-coated cells?

(A) C1 through C3
(B) C1 through C5
(C) C1 through C6
(D) C1 through C8
(E) C1 through C9

422. Which of the following terms could be used to describe the trans-plantation of a kidney from one identical twin to the other twin?

(A) Autograft
(B) Isograft
(C) Homograft
(D) Heterograft
(E) Xenograft

423. The Arthus reaction is a classic inflammatory response that best is described by which of the following statements?

(A) The Arthus reaction requires a low concentration of antigen and antibody
(B) The Arthus reaction appears later after injection than does passive cutaneous anaphylaxis
(C) The Arthus reaction is mediated by immunoglobulin M
(D) The characteristic Arthus lesion develops slowly
(E) The extent of the Arthus lesion is independent of the quantity of reacting antigen and antibody

424. The immunodeficiency associated with chronic granulomatous disease is related to

(A) the absence of the third component of complement (C3)
(B) the reduction in the levels of the fifth component of complement (C5)
(C) the inability of polymorphonuclear leukocytes to ingest bacteria
(D) the inability of polymorphonuclear leukocytes to kill ingested bacteria
(E) dysgammaglobulinemia

425. Which of the following serum immunoglobulin (Ig) classes is characteristically elevated in a newborn infant who had had an in-utero infection?

(A) IgA
(B) IgD
(C) IgE
(D) IgG
(E) IgM

426. The light and heavy chains of immunoglobulins M and G are joined by

(A) carbon-carbon double bonds
(B) peptide bonds
(C) disulfide bonds
(D) side-chain ester linkages
(E) glycosidic linkages

427. The presence of agglutinating *Brucella* antibodies (IgG) at a titer of 1:80 or greater is indicative of

(A) *Brucella* vaccination
(B) *Brucella* carrier state
(C) active brucellosis
(D) chronic brucellosis
(E) active leptospirosis

428. The Venereal Disease Research Laboratories (VDRL) test is a common screening test for syphilis. The immunologic basis for this test is the presence of

(A) serum antibodies to surface antigens of *Treponema pallidum*
(B) serum antibody to a cardiolipin that cross-reacts with *Treponema* antigen
(C) serum antibody that cross-reacts with C-reactive protein (CRP)
(D) whole *Treponema* organisms
(E) soluble *Treponema* antigen

429. The passive transfer of cutaneous anaphylaxis in humans is known as the Prausnitz-Küstner reaction. This reaction requires

(A) time for the antigen to be altered by macrophages
(B) time for the antigen to be deposited in subcutaneous venules
(C) time for the antigen to fix to skin receptors
(D) time for the antigen to be enzymatically activated
(E) none of the above

430. Peyer's patches are aggregated lymphatic follicles present in largest number in the

(A) duodenum
(B) jejunum
(C) ileum
(D) cecum
(E) colon

431. Which of the following classes of immunoglobulin is transferred passively from mother to fetus?

(A) IgA
(B) IgD
(C) IgE
(D) IgG
(E) IgM

432. Testing during anaphylaxis for which of the following substances would most likely be NEGATIVE?

(A) Histamine
(B) Kinin
(C) Pyrogen
(D) Serotonin
(E) Slow reacting substance (SRS-A)

433. Rheumatoid factor is elevated in individuals afflicted with rheumatoid arthritis. Chemically, rheumatoid factor is

(A) a complement component
(B) an HL-A-linked serum globulin
(C) a C-reactive protein
(D) denatured serum albumin
(E) an immunoglobulin

434. The hinge region of an IgG heavy chain is located

(A) between V_H and C_{H1}
(B) between C_{H1} and C_{H2}
(C) between C_{H2} and C_{H3}
(D) within the C_{H1} intrachain disulfide loop
(E) within the Fc fragment

435. Which of the following statements about T cells is NOT true?

(A) The majority of circulating lymphocytes are T cells
(B) T cells have far fewer surface immunoglobulins than do B cells
(C) T cells are distributed in interfollicular areas of lymph nodes
(D) T cells proliferate in response to an antigen to which they have been primed
(E) T cells differentiate into antibody-secreting plasma cells

436. Which of the following subclasses of antibody is most abundant in serum?

(A) IgG-1
(B) IgG-2
(C) IgG-3
(D) IgG-4
(E) IgA-1

437. The class or subclass of antibody that sensitizes human mast cells for anaphylaxis is

(A) IgG-1
(B) IgG-3
(C) IgA-2
(D) IgM
(E) IgE

438. Genes that govern the ability of an animal to react to a given antigen are called

(A) HL-A genes
(B) immune response genes
(C) immunoglobulin genes
(D) allogenes
(E) oncogenes

439. In humans, two closely linked genetic loci, each made up of two alleles, comprise the histocompatibility locus A (HL-A). Paired first and second locus antigens are called haplotypes. The HL-A haplotypes (separated by a semicolon) of a child's parents are given below. Assuming that no crossover events have occurred, the child's histotype could be which of the following?

Father 3,25; 7,12
Mother 1,3; 8,9

(A) 1,3; 7,8
(B) 7,12; 1,3
(C) 3,3; 7,9
(D) 1,25; 7,12
(E) 3,25; 7,12

440. Many autoimmune and oncologic diseases show a statistical association with the locus for histocompatibility antigen. The correlation is thought to occur because HL-A genes and other closely placed genes in the chromosome do not segregate randomly. Disease susceptibility, while loosely correlated with HL-A type, is felt to be more closely associated with another genetic locus, which is the

(A) Z gene for alpha$_1$-antitrypsin
(B) alpha immunoregulatory peptide gene
(C) immune response (Ir) genes
(D) immunoglobulin genes
(E) secretor-nonsecretor genes

441. The amounts of protein precipitated in a series of tubes containing a constant amount of antibody and varying amounts of antigen are presented below. In which tube is antigen-antibody equivalence obtained?

	Tube	Antigen (mg)	Protein precipitated (mg)
(A)	1	0.02	1.1
(B)	2	0.08	2.1
(C)	3	0.32	3.1
(D)	4	1.0	3.7
(E)	5	2.0	2.9

442. Alleles of two or more linked genes occur together in a population more frequently than would be expected by random segregation. This phenomenon is called

(A) allotypic dominance
(B) codominant linkage
(C) hysteresis
(D) linkage disequilibrium
(E) nondisjunction

443. Most of the white blood cells found in the peripheral blood of adults are

(A) granulocytes
(B) macrophages
(C) null lymphocytes
(D) B lymphocytes
(E) T lymphocytes

444. Thrombocytopenic individuals who have aplastic anemia benefit from platelet transfusions during periods of severe platelet depression. Platelets may be rejected due to an ABO or HL-A incompatibility. Which of the following platelet-transfusion donors would be best for a patient whose blood type is AB+ and HL-A haplotypes 1,3; 7,12?

(A) O+ 1,3; 7,12
(B) A+ 3,5; 7,8
(C) O+ 3,5; 7,8
(D) AB+ 2,−; 5,13
(E) O− 1,−; 7,13

445. The principal deterrent to successful transplantation of human hearts is which of the following?

(A) High infection rate
(B) Inability to control rejection
(C) Psychosocial objections
(D) Unreliable surgical technique
(E) Unavailability of donor hearts

446. The most commonly transplanted organ in humans is

(A) bone marrow
(B) kidney
(C) liver
(D) pancreas
(E) skin

447. Which of the following cell surface alloantigens is NOT expressed by mouse T cells?

(A) H-2
(B) Ly
(C) PC
(D) Theta
(E) TL

448. An Ouchterlony gel diffusion plate shows the reaction of a polyspecific serum against several antigen preparations. The center well in Figure 1 below contains polyspecific antiserum, first bleed; the center well in Figure 2 contains polyspecific antiserum, second bleed; NS is normal saline. In this situation, cross reaction can be recognized between antigen X and

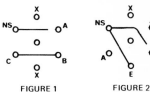

FIGURE 1   FIGURE 2

(A) antigen A
(B) antigen B
(C) antigen C
(D) antigen D
(E) antigen E

449. Transfer factor has the capacity to transfer delayed hypersensitivity to another, nonreactive individual. Transfer factor is best described by which of the following statements?

(A) It is not destroyed by DNase or RNase
(B) It is trypsin-sensitive
(C) It is immunologically nonspecific
(D) It is derived from erythrocytes
(E) Its molecular weight is greater than 200,000

Questions 450-452

The Scatchard plot shown below represents the interaction of a hapten molecule with an immunoglobulin in an equilibrium dialysis apparatus. This interaction is defined by the equation

$$K=\left(\frac{r}{n-r}\right)c \quad \text{or} \quad \frac{r}{c}=Kn-Kr$$

(K is the intrinsic affinity constant, c is the free concentration of hapten, r is the number of hapten molecules bound per antibody molecule at c, and n is the antibody valence.)

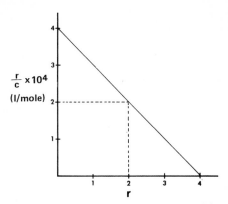

450. The affinity constant for this system is

(A) 1×10^{-4} moles/l
(B) 1×10^4 moles/l
(C) 1×10^4 l/mole
(D) -4×10^{-4} moles/l
(E) 16×10^{-4} moles/l

451. The antibody valence, n, is defined as the maximum number of ligand molecules able to be bound per antibody molecule. In the example presented, n equals

(A) 1
(B) 2
(C) 3
(D) 4
(E) 10

452. The antibody species most likely to have been used in the experiment described is

(A) IgA
(B) IgD
(C) IgE
(D) IgG
(E) IgM

453. A test is characterized by a serial dilution of a patient's serum, which is incubated with antigen and to which washed erythrocytes are added. Results are reported in Todd units. This test is designed for the detection of

(A) antinuclear antibodies
(B) macroglobulins
(C) cryoglobulins
(D) C-polysaccharides
(E) antistreptolysin O

454. The C-reactive protein (CRP) test, which is used in the study of inflammatory diseases, detects the presence of C substance, which can be best described as a

(A) human cold agglutinin
(B) Reiter-strain treponeme
(C) *Proteus* OX-19 flagellar antigen
(D) pneumococcal cell-wall antigen
(E) *Haemophilus* capsular antigen

DIRECTIONS: Each question below contains four suggested answers of which **one** or **more** is correct. Choose the answer:

A	if	**1, 2, and 3**	are correct
B	if	**1 and 3**	are correct
C	if	**2 and 4**	are correct
D	if	**4**	is correct
E	if	**1, 2, 3, and 4**	are correct

455. Removal of the bursa of Fabricius from a chicken results in

(1) a markedly decreased circulating-lymphocyte count
(2) atrophy of the germinal centers of the spleen
(3) delayed rejection of skin grafts
(4) a drop in serum levels of immunoglobulin G

456. The Ouchterlony gel diffusion pattern shown below represents reactions of various fractions in the preparation of a membrane component X with antisera against the crude membrane. In the diagram, the center well contains antimembrane antiserum; NS is normal saline. Component X is found in

(1) well 1
(2) well 2
(3) well 3
(4) well 4

457. Which of the following physiologic disturbances have been observed in anaphylaxis?

(1) Edema
(2) Release of heparin
(3) Leukopenia
(4) Smooth-muscle spasm

458. Among the nonspecific laboratory data that may aid in the diagnosis of infectious disease are

(1) serum level of C-reactive protein
(2) white blood cell count
(3) serum level of lactate dehydrogenase
(4) erythrocyte sedimentation rate

459. Antibody production is dependent upon the

(1) amount of antigen administered
(2) route of administration of the antigen
(3) time period over which the antigen is administered
(4) inherent antigenicity of the antigen

SUMMARY OF DIRECTIONS

A	B	C	D	E
1, 2, 3 only	1, 3 only	2, 4 only	4 only	All are correct

460. The Arthus reaction differs from the reaction of anaphylaxis in that the Arthus reaction

(1) may be delayed in development for several hours
(2) has a latent period for fixation of antibody to tissue cells
(3) involves complement fixation
(4) is inhibited by antihistaminic agents

461. Autoimmunity to sequestered protein has been implicated strongly in which of the following diseases?

(1) Thrombotic thrombocytopenic purpura
(2) Goodpasture's syndrome
(3) Rheumatic fever
(4) Hashimoto's thyroiditis

462. Which of the following statements about mouse B lymphocytes are true?

(1) They have cell-surface immunoglobulins
(2) They are more common than T lymphocytes in peripheral blood
(3) They are more likely than T lymphocytes to be inactivated by X-irradiation
(4) They have theta-antigen cell-surface markers

463. The graph below shows the sequential alteration in the type and amount of antibody produced after an immunization. (Inoculation of antigen occurs at two different times, as indicated by the arrows.) Curve A and curve B each represents a distinct type of antibody. The class of immunoglobulin represented by curve B is

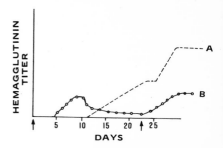

(1) estimated to have a molecular weight of 150,000 daltons
(2) composed of four peptide chains connected by disulfide links
(3) not produced in neonates until approximately the third month of life
(4) the human ABO isoagglutinin

464. Kappa and lambda light chains have which of the following characteristics?

(1) They are identical at the Inv locus
(2) They both are precursors of Bence Jones proteins
(3) They both are found on the same heavy chain dimer
(4) The genes in which they are coded are on different chromosomes

465. Which of the following statements about the precipitin curve shown below are true?

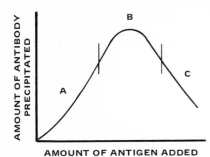

(1) In a multispecific system, a solution in zone "B" may have an excess of antigen and antibody in the supernatant
(2) A solution in zone "C" would be expected to have an excess of antibody in the supernatant
(3) In a monospecific system, a solution in zone "B" would contain only reacted antibody and antigen
(4) A solution in zone "A" would be expected to have unreacted precipitable antigen in the supernatant

466. J chains are found in which of the following immunoglobulins?

(1) IgG
(2) IgM
(3) IgE
(4) IgA

467. Which of the following antigen-antibody combinations would form precipitin bands between wells in a gel diffusion experiment?

(1) Anti-Fab, heavy chains
(2) Anti-Fab, light chains
(3) Anti-Fc, heavy chains
(4) Anti-Fc, light chains

468. Which of the following hypotheses would sufficiently explain non-precipitation in an antigen-antibody system?

(1) The antigen has a monovalent determinant
(2) The antigen has multiple, closely repeated determinants
(3) The antibody has been cleaved to divalent Fab' ligands
(4) The antibody has been cleaved to divalent $F(ab')_2$ ligands

469. There are two main classes of lymphocytes: B lymphocytes and T lymphocytes. B and T cells may be differentiated by

(1) a determination of surface receptors for C3b
(2) response to plant mitogens
(3) the presence of a surface allo-antigen
(4) the capacity for antigenic memory

470. Immune complex glomerulonephritis is seen in which of the following conditions?

(1) Chronic serum sickness
(2) Phenacetin ingestion
(3) Systemic lupus erythematosus
(4) Hemolytic-uremic syndrome

SUMMARY OF DIRECTIONS

A	B	C	D	E
1, 2, 3 only	1, 3 only	2, 4 only	4 only	All are correct

471. A rabbit is injected with *p*-azobenzenearsonate and fails to produce antibodies. Inoculation of benzenearsonate coupled to a nonspecific protein induces the rabbit to produce antibodies to the benzene-arsonate-protein complex. In this situation, benzenearsonate acts as a "hapten"; haptens are

(1) generally of low molecular weight (<1,000 daltons)
(2) able to elicit an immune response usually only if coupled covalently to a carrier
(3) bound to antigen-binding sites of immunoglobulins elaborated in response to the hapten-carrier complex
(4) coupled to endogenous proteins prior to allergic sensitization

472. Which of the following procedures can yield a relatively pure population of T lymphocytes from peripheral blood?

(1) Sedimentation through a ficoll-hypaque density gradient after incubation and rosette formation with sheep erythrocytes
(2) Separation on a cell separator by applying an electrostatic charge and field to drops with fluorescein-tagged anti-IgG and collecting the pool of nontagged cells
(3) Passage through nylon fiber columns, which bind IgG-bearing B cells and macrophages
(4) Treatment with anti-theta sera and incubation for 30 minutes with complement

473. Allotypes have been identified for which of the following molecules?

(1) Glucose-6-phosphate dehydrogenase
(2) Haptoglobin
(3) Hemoglobin
(4) HL-A histocompatibility antigen

474. Relative to the primary immunologic response, secondary and later
booster responses to a given hapten-
protein complex can be associated with

(1) higher titers of antibody
(2) increased antibody affinity for
    the hapten
(3) increased antibody avidity for the
    original hapten-protein complex
(4) a shift in subclass or idiotype of
    antibody produced

475. False-positive reactions in serologic tests for syphilis are known to
occur if tested individuals suffer from

(1) leprosy
(2) malaria
(3) disseminated lupus erythematosus
(4) periarteritis nodosa

DIRECTIONS: The groups of questions below consist of lettered choices followed by several numbered items. For each numbered item select the **one** lettered choice with which it is **most** closely associated. Each lettered choice may be used once, more than once, or not at all.

Questions 476-480

For each type of immunoglobulin (Ig) below, select the appropriate molecular weight.

(A) 150,000 (7S)
(B) 180,000 (7S)
(C) 190,000 (8S)
(D) 400,000 (11S)
(E) 900,000 (19S)

476. IgA (secretory)

477. IgD (serum)

478. IgE

479. IgG

480. IgM (serum)

Questions 481-485

For each disease listed below, choose the level of immune function (humoral and cellular) with which it is most likely to be associated.

	Humoral	**Cellular**
(A)	Normal	Normal
(B)	Normal	Deficient
(C)	Deficient	Normal
(D)	Deficient	Deficient
(E)	Elevated	Elevated

481. Ataxia-telangiectasia

482. Infantile X-linked agammaglobulinemia (Bruton's disease)

483. Swiss-type hypogammaglobulinemia

484. Thymic hypoplasia (DiGeorge's syndrome)

485. Wiskott-Aldrich syndrome

Questions 486-489

For each technique or phenomenon described below, select the investigator with whom it is associated.

(A) Coombs
(B) Danysz
(C) Farr
(D) Karush
(E) Ouchterlony

486. Precipitation of antigen-antibody complexes from a high-salt solution in which the uncomplexed antigen is soluble

487. Indirect agglutination of human red blood cells by rabbit anti-human IgG, which recognizes human antibodies that are adsorbed onto red blood cells but that are, themselves, nonagglutinating

488. The serial addition over a period of time of several portions of antigen (toxin) to an antibody (antitoxin), yielding a different equivalence point from that obtained when the entire amount of antigen is added to the same amount of antibody all at once

489. Nonprecipitation of antigen and antibody, because the antigen contains closely spaced repetitive antigenic determinants and because both arms of the antibody bind determinants on the same antigenic particle (i.e., monogamous bivalency)

Questions 490-492

Antigenic determinants on immunoglobulins are used to classify antibodies. For each antibody classification below, select the determinant with which it is most likely to be associated.

(A) Determinant exposed after papain cleavage to an $F(ab')_2$ fragment
(B) Determinant from one clone of cells and probably located close to the antigen-binding site of the immunoglobulin
(C) Determinant inherited in a Mendelian fashion and recognized by cross-immunization of individuals in a species
(D) Heavy-chain determinant recognized by heterologous antisera
(E) Species-specific carbohydrate determinant on the heavy chain

490. Isotype

491. Allotype

492. Idiotype

Questions 493-497

For each diagnosis given, choose the serum electrophoretic profile with which it is most likely to be associated.

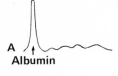

A
Albumin

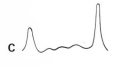

B

C

D

E

Questions 498-500

Complement-fixation (CF) testing is an important serologic tool. For each reaction mixture below, select the expected result.

(A) Complement is bound, red blood cells are lysed
(B) Complement is bound, red blood cells are not lysed
(C) Complement is not bound, red blood cells are lysed
(D) Complement is not bound, red blood cells are not lysed
(E) Complement is not bound, red blood cells are agglutinated

498. Anti-*Mycoplasma* antibody + complement + hemolysin-sensitized red blood cells (RBC) + anti-RBC antibody

499. Anti-*Mycoplasma* antibody + *Mycoplasma* antigen + complement + hemolysin-sensitized red blood cells

500. Anti-*Mycoplasma* antibody + *Mycoplasma* antigen + complement + hemolysin-sensitized red blood cells + anti-RBC antibody

493. Alpha$_1$-antitrypsin deficiency

494. Multiple myeloma

495. Swiss-type agammaglobulinemia

496. Polyclonal hypergammaglobu-linemia

497. Normal

Immunology

Answers

412. The answer is A. *(Davis, ed 2. pp 456-461, 572-573.)* Antibody (e.g., antistreptolysin O) is formed and secreted both by plasma cells and by B cells, which are lymphocytes derived from bone marrow. B cells are responsible for humoral immunity. Thymus-derived lymphocytes (T cells) are associated with cell-mediated immunity and elaborate many factors, including lymphotoxin, blastogenic factor, migration-inhibition factor, transfer factor, as well as a number of stimulatory and suppressive factors that are less well defined. Thus, the T cell is considered the regulator cell of the immune system.

413. The answer is D. *(Bellanti, p 131.)* Each immunoglobulin class has a biochemically different type of heavy chain; this property is responsible for antigenic differences observed among the classes. The light chains, on the other hand, are common to the various immunoglobulin classes. Secretory component is a fragment added to immunoglobulin A (IgA) in specialized epithelial cells. The J chain is a component of IgM and IgA multimers.

414. The answer is C. *(Bellanti, p 631.)* Bence Jones proteins are homogeneous free-globulin light chains (kappa and lambda chains) present in the urine of about half of all persons with multiple myeloma. Because Bence Jones protein is not albumin, "dipstick" reagents often employed to monitor urine for "protein" (i.e., albumin) are ineffective. Mu and gamma chains are types of heavy chains.

415. The answer is D. *(Davis, ed 2. pp 506-507.)* Recurrent severe infection is an indication for clinical evaluation of immune status. Live vaccines, including BCG attenuated from *Mycobacterium tuberculosis*, should **not** be used in the evaulation of an individual's immune competence, because individuals with severe immunodeficiencies may develop an overwhelming infection from the vaccine. For the same reason, oral (Sabin) polio vaccine is not advisable for use in such persons.

416. The answer is D. *(Davis, ed 2. pp 1346-1347. Jawetz, ed 13. pp 422-423.)* Measles virus, or a virus identical to it, has been isolated from the lymph nodes and brains of individuals who have died of subacute sclerosing panencephalitis (SSPE). Immunofluorescence studies have demonstrated antigens reactive with measles antibody, and patients with the disease have high titers of measles antibody. SSPE, which primarily affects children and adolescents, is a progressive, degenerative disease of the white matter underlying the cortex of the brain. The presence of latent intracellular measles virus suggests a defect in an affected individual's cellular immunity; persistent viremia eventually permits measles-virus-laden lymphocytes to invade the central nervous system.

417. The answer is E. *(Davis, ed 2. pp 1314-1317.)* The surface hemagglutinin, which corresponds to the surface "spikes" seen by electron microscopy, is the major target of neutralizing antibody against influenza virus. Antibody against neuraminidase does not prevent infection, although it reduces viral spread from infected cells. Antibodies formed against the nucleocapsid, nucleoprotein, and M protein do not neutralize the infectivity of influenza virus. M protein and nucleocapsid antigens provide the basis for the typing of influenza virus. *not according to our class*

RNP - types   H & N spikes - Subtypes

418. The answer is A. *(Gell, ed 3. p 135.)* Virus neutralization tests can detect as little as 0.0001 mg of antibody nitrogen per test. Complement fixation and radioimmunoassay also are highly sensitive, though less so than neutralization. Less sensitive still are immunodiffusion and immunoelectrophoretic assays.

419. The answer is A. *(Bellanti, p 116.)* IgA is the principal immunoglobulin in exocrine secretions. Consequently, it provides most of the humoral protection against pathogens invading the mucosal surfaces of the respiratory, intestinal, and genitourinary tracts. IgA is a multimeric antibody that contains a "secretory piece" elaborated by specialized epithelial cells.

420. The answer is B. *(Davis, ed 2. pp 546-551, 638.)* Inactivated vaccine stimulates the formation of high levels of serum antibodies that form complexes with live viruses. The resulting accumulation of humoral and cellular factors mediates the immunologic tissue injury of the Arthus reaction. The Prausnitz-Küstner reaction concerns passively transferred antibody later challenged intradermally with antigen. The Shwartzman reaction is a two-step cutaneous response to bacterial endotoxin. The Schick reaction, elicited with intradermal diphtheria toxin, is used to assay immunity to diphtheria. The Coombs reaction is an in-vitro antiglobulin test to demonstrate the presence of erythrocytic antibodies.

421. The answer is E. *(Davis, ed 2. pp 517-519.)* All nine components of complement must interact to produce swift cell lysis. However, if all components but C9 are bound, functional membrane lesions develop, which on incubation at 37°C may lead to slow lysis. Subsequent binding of C9 to this complex accelerates lysis and produces circular membrane lesions similar to those produced by detergent polyene antibiotics.

422. The answer is B. *(Davis, ed 2. p 610.)* Isografts are grafts between genetically identical individuals. Autografts are transplants of an individual's own tissue from one region to another. Homografts are transplants from a genetically nonidentical individual of the same species. Heterografts (or xenografts) are transplants from one species to another.

423. The answer is B. *(Bellanti, p 310. Davis, ed 2. pp 538-551.)* The Arthus reaction is a classic inflammatory response involving a cellular infiltrate provoked by antigen and antibody in much larger quantities than those required for passive cutaneous anaphylaxis. Whereas the edema of cutaneous anaphylaxis appears within 10 minutes and resolves within 30 minutes of antigen injection, the polymorphonuclear leukocyte infiltrate of an Arthus reaction appears after more than an hour, peaks at three to four hours, and resolves within 12 hours. The severity of the Arthus reaction is proportional to the amount of antigen and antibody reacting. With high antigen-antibody concentrations, necrosis may result.

424. The answer is D. *(Bellanti, p 684.)* A metabolic abnormality in the leukocytes of individuals with chronic granulomatous disease leads to defective intracellular killing of catalase-positive bacteria of low virulence. Leukocytes in affected patients fail to exhibit a postphagocytic burst of oxygen consumption and glucose metabolism and appear to lack NADPH oxidase activity. Ultimately, they fail to produce hydrogen peroxide and compromise myeloperoxidase-halogenation mechanisms used to kill bacteria in leukocytes.

425. The answer is E. *(Bellanti, pp 70-73.)* Until the second or third month postpartum, the gamma globulin in the blood of an infant is predominantly immunoglobulin G (IgG) passively acquired from the mother. However, a fetus who develops an infection responds by producing antibody of the IgM class, which is ontogenically the earliest form of antibody. Because IgM cannot traverse the placenta, presence of IgM in a neonate is a sign of past fetal infection.

426. The answer is C. *(Bellanti, pp 119-121.)* The heavy and light chains of immunoglobulins M and G are linked covalently by disulfide bonds. The two heavy chains in IgG molecules also are linked by disulfide bonds, as are the five subunits of IgM. Disulfide bonds can be broken by mild reduction.

427. The answer is C. *(Jawetz, ed 13. p 291.)* It may be difficult and time-consuming to isolate *Brucella* from individuals who have brucellosis. Serologic detection of the disease, then, becomes significant. IgG antibodies appear during the acute stages of the disease; IgM antibodies may persist for years.

428. The answer is B. *(Jawetz, ed 13. p 292.)* Serologic tests for syphilis, such as the VDRL test, are based on the fortuitous cross-reaction between *Treponema pallidum* antigen (reagin) and certain lipid fractions of mammalian tissue. These tests are sensitive but not specific—many disorders, including the so-called collagen diseases, give false positive results. A more specific test for syphilis is the fluorescent treponemal antibody-absorption (FTA-ABS) test.

429. The answer is C. *(Davis, ed 2. p 530.)* The Prausnitz-Küstner (P-K) reaction classically requires a latent period of 10 to 20 hours for sensitization to occur. This period of time allows antigen to become fixed to skin receptors. The antigen can elicit a wheal-and-flare response for as long as six weeks after injection of antibody. The P-K reaction is used to test individuals for allergic response to fungi, food products, plant pollens, animal danders, and other allergens.

430. The answer is C. *(Davis, ed 2. p 477.)* Gut-associated lymphatic aggregates (lymphoepithelial structures) include the tonsils, appendix, and Peyer's patches. Follicular organization in these structures follows the same pattern: B-cell preponderance in germinal centers and T-cell preponderance in the intervening parafollicular zones. Peyer's patches are found chiefly in the distal ileum. Their inflammation in children, following a viral illness, can lead to the condition known as intussusception.

431. The answer is D. *(Bellanti, pp 70-73.)* IgG is the only immunoglobulin to cross the placenta. Its molecular weight of 150,000 is less than that of IgM (900,000) and IgA (170,000 or 400,000), which exist chiefly as multimers and are too large to pass through the vasculature of the placental membranes. IgD and IgE, which are monomeric immunoglobulins (molecular weights 180,000 and 190,000, respectively) found normally in minute quantities in serum, are not known to pass through the placenta.

432. The answer is C. *(Bellanti, pp 301-302.)* Histamine is released primarily from mast cells and serotonin from platelets during anaphylaxis. Kinins, especially bradykinin, are increased during anaphylaxis. Slow reacting substance (SRS-A) is an acidic lipid released in anaphylaxis, probably from the lung. Although some prostaglandins and other mediators appear to have roles in anaphylaxis, endogenous pyrogen does not appear to be one of these; anaphylaxis is associated more often with hypotension, hypothermia, and cool shock than with hyperthermia.

433. The answer is E. *(Davis, ed 2. p 419.)* Rheumatoid factors are immunoglobulins, usually IgM and less often IgG, in the sera of individuals who have rheumatoid arthritis. These factors can react with red blood cells or other particles coated with normally occurring human IgG. An autoimmune mechanism has been postulated by some investigators both to explain the existence of rheumatoid factors and to elucidate the cause of rheumatoid arthritis.

434. The answer is B. *(Davis, ed 2. p 429.)* The hinge region of immunoglobulin G exists between C_{H1} and C_{H2}, as illustrated below.

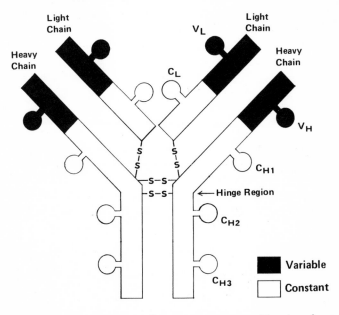

C regions are polypeptide segments of constant sequence; V regions have a variable amino-acid sequence. They occur in both the heavy (H) and light (L) chains. Differences in polypeptide sequence undoubtedly contribute to the distinctive biologic properties of the various immunoglobulins.

435. The answer is E. *(Davis, ed 2. pp 459-460, 466.)* Most circulating lymphocytes in the human body are T cells. In lymph nodes, T cells are located in the interfollicular regions. T cells are associated with cell-mediated immunity and proliferate in response to specific antigens. B cells, not T cells, differentiate into plasma cells; however, T cells help in this process. B cells have far more surface antibody than found on T cells.

436. The answer is A. *(Davis, ed 2. pp 406-407, 446.)* The serum concentrations (mg/ml) for each of the subclasses of immunoglobulins listed in the question are as follows: IgG-1, 8 mg/ml; IgG-2, 4 mg/ml; IgG-3, 1 mg/ml; IgG-4, 0.4 mg/ml; and IgA-1, 3.5 mg/ml. Serum concentrations of other antibodies are: IgA-2, 0.4 mg/ml; IgM, 1 mg/ml; IgD, 0.03 mg/ml; and IgE, 0.0001 mg/ml. Subclasses of IgG differ in several respects, including effector function and, to a lesser degree, antigenic characteristics.

437. The answer is E. *(Davis, ed 2. pp 446, 538.)* Immunoglobulin E molecules sensitize human mast cells and are responsible for atopic allergy. In human sera, IgE is normally present in a concentration that is 0.004% of the IgG concentration. IgE-producing cells are located in the mucosa of the respiratory and gastrointestinal tracts.

438. The answer is B. *(Davis, ed 2. pp 477-479.)* In the mouse, immune response genes are located between genes coding for serologically defined K and D antigens of the H-2 histocompatibility complex. The genes govern the ability of a mouse to make circulating antibodies in response to booster injections of "limited" antigens, i.e., antigens that (1) are given in small doses, (2) possess few structural determinants, and (3) have only a few structural determinants that differ from the test antigen (e.g., natural alloantigens such as the histocompatibility antigens).

439. The answer is B. *(Davis, ed 2. p 618.)* In the question presented, the haplotypes of the father are 3,25 and 7,12 and the haplotypes of the mother are 1,3 and 8,9. (A haplotype is composed of one allele—antigen—from one gene of a pair and one allele from the other gene.) Each child of this couple would have inherited one haplotype from each parent. Thus, possible offspring haplotypes are (i) 3,25; 1,3; (ii) 3,25; 8,9; (iii) 7,12; 1,3; and (iv) 7,12; 8,9.

440. The answer is C. *(Davis, ed 2. pp 478-480.)* Experiments using mice have shown that a strain's immune response to specific antigens is an ability closely linked to that strain's histocompatibility antigens. More recently, investigators have uncovered the close proximity between histocompatibility genes and immune response (Ir) genes; indeed, disease susceptibility now is thought to be linked more closely to the immune response genes than to the histocompatibility genes themselves. The analysis of human histocompatibility gene linkage to human immune response genes has lagged far behind the analysis of the murine model.

441. The answer is D. *(Davis, ed 2. pp 371-373.)* In tube four in the question presented, the maximum protein precipitate is observed. According to the rules governing precipitin reactions, maximum precipitation occurs at approximately antigen-antibody equivalence. In tubes one through three, antibody excess occurs; in tube five, antigen excess occurs.

442. The answer is D. *(Davis, ed 2. pp 233, 420, 616-618.)* Linkage disequilibrium occurs when the concurrent expression of alleles of two loci happens more often than could be statistically predicted. Presumably, there is a selective advantage to the expression of both loci together. Genetic linkage — i.e., the relative positioning of different loci — is determined by the frequency with which these loci recombine during a genetic cross.

443. The answer is A. *(Davis, ed 2. p 644.)* Granulocytes constitute about 70 percent of all white blood cells in peripheral blood smears from adults. Macrophages rarely are seen. Of the lymphocytes, T cells constitute 70 percent, B cells approximately 25 percent, and null (neither T nor B) cells about 5 percent. In children, lymphocytes may predominate in the peripheral smear.

444. The answer is A. *(Davis, ed 2. pp 618-620.)* From the point of view of ABO compatibility, an AB+ individual would not be expected to reject any of the types of transfused platelets listed in the question. (Similarly, an Rh+ individual should be able to accept Rh− blood elements.) Thus, the choice of donor would be based, in this instance, on HL-A compatibility, so that the O+, HL-A 1,3; 7,12 donor would be the best for the patient described.

445. The answer is B. *(Davis, ed 2. pp 610-614.)* Surgical techniques for transplanting human hearts are well developed. However, the exceedingly difficult problem of rejection of transplanted human organs has yet to be overcome. Intensive immunosuppression with steroids has proven helpful but not fully rewarding.

446. The answer is E. *(Davis, ed 2. pp 610-614. Gell, ed 3. p 511.)* Autografting of skin is the most common surgical transplantation in humans. Skin autografts are employed to cover surface areas denuded by burns, severe lacerating trauma, operative procedures (e.g., radical mastectomy on occasion), and other causes. Allografts can be used as life-saving emergency measures to control fluid loss and infection in some severely injured individuals; unlike autografts, however, allografts are subject to tissue rejection.

447. The answer is C. *(Davis, ed 2. p 459.)* Although mouse T lymphocytes do not express the PC (plasma cell) surface alloantigen, they, like human T cells, do express histocompatibility (H-2) antigens. The Ly and TL alloantigen systems, which have been intricately described, allow the functional subclasses of the T-cell population to be differentiated. Theta antigen, shared by brain and T cells, is an analog of T-cell antigens in other species, which may be defined by isoimmunization.

448. The answer is A. *(Davis, ed 2. pp 387-389.)* In the Ouchterlony agar-gel diffusion test, an antigen and a series of antibodies (or, an antibody and a series of antigens) are allowed to diffuse toward each other. At the zone of optimal proportions of the reactants, a precipitin line occurs. Cross reactions between antigens or antibodies tested can be detected by 1) a shortening of the major precipitin band contiguous to the cross-reacting substance or 2) the identity of precipitin reaction between the two cross-reacting substances. The figures presented in the question illustrate both types of cross reaction. In the first bleed pattern shown in the question, cross reaction between antigen X and antigen A is recognizable only by the shortening of the precipitin band between the center well and X on the A well side (relative to the band going directly into the normal saline well). In the second bleed pattern, full cross reaction of X and A is apparent. No other cross reactions are seen.

449. The answer is A. *(Bellanti, p 235.)* Transfer factor is an immunologically specific substance that is resistant to nucleases and proteases. It is derived from leukocytes and has a molecular weight of less than 10,000 (perhaps even less than 5,000). Transfer factor appears to mediate the human cellular immune response; its injection into individuals whose cellular immunity is deficient confers temporary immune competence.

450. The answer is C. *(Davis, ed 2. pp 361-362.)* In a Scatchard plot, the slope of the line is equal to $-K$. As shown in the diagram presented, the slope ($r/c \div r$) is $-[(2 \times 10^4 \text{ l/mole}) \div 2] = -1 \times 10^4$ l/mole. Thus, K equals 1×10^4 l/mole.

451. The answer is D. *(Davis, ed 2. pp 361-362.)* In the graph presented, as r approaches 4, r/c approaches 0 and, consequently, c approaches infinity. In general, the x-intercept (i.e., $r/c = 0$) gives the number of ligand binding sites at maximal saturation. In the example described, this maximum number of ligand molecules able to be bound per antibody molecule — or, the antibody valence — is four. Antibody valence also can be calculated from the equation given; i.e., when $r/c = 2$ x 10^4 1/mole, $r = 2$, and $K = 1$ x 10^4 1/mole, then $n = [(r/c + Kr) \div K]$ = $[(2$ x $10^4 1/mole + 2$ x $10^4 1/mole) \div 1$ x $10^4 1/mole] = 4$.

452. The answer is A. *(Davis, ed 2. pp 410, 417-418.)* Secretory IgA is a tetravalent dimer and thus would have an antibody valence of 4 (as calculated in the preceding question). IgG and IgE are divalent immunoglobulins. IgM is pentavalent or decavalent, depending on the experimental conditions.

453. The answer is E. *(Finegold, ed 5. pp 414-415.)* A rising titer of antistreptolysin O (ASLO) is detected in the serum of patients having had a recent infection with group A streptococci. Commercially available streptolysin O is added to serial dilutions of patient's serum. Human or rabbit erythrocytes are added to act as an indicator of whether or not the patient's antibodies inactivated the added streptolysin O, thus inhibiting hemolysis. The ASLO titer is reported in Todd units.

454. The answer is D. *(Finegold, ed 5. p 415.)* The antigen derived from pneumococcal cell walls and composed chiefly of polysaccharides is called C substance. C-reactive protein (CRP) is an abnormal protein occurring in the serum of many patients having inflammatory diseases. Testing for CRP is, therefore, a nonspecific procedure of little diagnostic value. It is useful, however, as an indicator of the course of a disease and "status" of the inflammation. More specific serologic methods using an anti-CRP serum have replaced the procedure of testing the reactivity of a patient's serum with the C-polysaccharide of the pneumococcal capsule.

455. The answer is C (2, 4). *(Davis, ed 2. p 474.)* Bursectomy inhibits the functioning of B lymphocytes. Germinal centers in the spleen atrophy, and the number of plasma cells as well as the serum level of circulating immunoglobulins decrease. T-cell functions are relatively unaffected by bursectomy; lymphocyte counts are maintained, and skin-graft rejection time is unaltered.

456. The answer is A (1, 2, 3). *(Davis, ed 2. p 389.)* In the Ouchterlony gel diffusion pattern depicted in the question, membrane component X cross-reacts with material in wells 1, 2, and 3. Component X, however, does not cross-react with the substance in well 4. A second component present in wells 1 and 2 does not cross-react with X or the material in well 4.

457. The answer is E (all). *(Davis, ed 2. pp 538-554.)* In anaphylaxis, edema occurs as a result of injury to vascular endothelium. Heparin is liberated, resulting in diminished coagulability of blood. Under the influence of several agents, including complement cascade fragments, leukocytes adhere to the walls of capillaries, especially in the lungs, and leukopenia results. Histamine release causes smooth-muscle spasm.

458. The answer is E (all). *(Jawetz, ed 13. pp 295-298.)* Certain nonspecific, acute-phase reactants, such as C-reactive protein, may be elevated in individuals who have an inflammatory illness. Similarly, valuable information on response to infection may be provided by the white blood cell count and differential and the erythrocyte sedimentation rate. When cells are injured by inflammatory processes, serum levels of intracellular enzymes like lactate dehydrogenase (LDH) may be elevated.

459. The answer is E (all). *(Jawetz, ed 13. pp 150-151.)* The antibody response to a given antigen depends upon the amount of antigen administered, the manner in which it is administered, the time period over which it is administered, and its inherent antigenicity. Adjuvants may heighten response to an antigen in cases where inherent antigenicity is less than optimal. Prior exposure of an individual to an antigen may prompt a secondary response to that antigen months or even years later.

460. The answer is B (1, 3). *(Jawetz, ed 13. pp 162-164.)* Anaphylactic reactions occur when specific cytotropic antibodies trigger the release of histamine from mast cells and basophils. When cell-bound antibodies react with antigen, degranulation and release of pharmacologic mediators (kinins, histamines) occur and lead to local or systemic reactions. In comparison, Arthus reactions require larger amounts of antibody, which binds with the appropriate antigen; in the process complement is fixed and polymorphonuclear leukocytes aggregate. The Arthus reaction is not inhibited by antihistamines.

461. The answer is D (4). *(Gell, ed 3. pp 1084, 1168-1170, 1218-1232, 1360-1375.)* Injury to tissues containing antigens that normally are not in contact with the body's immune mechanism (i.e., are "sequestered") can provoke a humoral and cellular autoimmune response; these antigens, in other words, are recognized as "foreign." In the case of Hashimoto's disease, thyroid antigens stimulate production of antithyroid antibody and cause infiltration of sensitized lymphocytes. In Goodpasture's syndrome, antibodies that react with basement-membrane material found in the lungs, kidneys, and elsewhere are generated following glomerulonephritis. In thrombocytopenic purpura, normal or drug-altered platelet membranes (not sequestered antigens) elicit and become the target of the host immune response. The antigen that elicits an immune response in individuals who have rheumatic fever is an antigen shared by group A beta-hemolytic streptococci and the human myocardial sarcolemma.

462. The answer is B (1, 3). *(Davis, ed 2. p 459.)* Of the lymphocytes in the peripheral blood of mice, B cells comprise 15 to 25 percent, while T cells make up 75 to 85 percent. B lymphocytes have cell-surface immunoglobulins, and T cells have theta cell-surface alloantigens. B cells are much more susceptible to X-irradiation than are T cells.

463. The answer is D (4). *(Davis, ed 2. p 484.)* The graph shown in the question exhibits hemagglutinating antibody responses to primary and secondary immunization with any standard antigen. Curve B represents the early response to primary immunization, which is chiefly an immunoglobulin M response. Rechallenge elicits an accelerated response that primarily involves immunoglobulin G and occurs two to five days after reimmunization. IgM has a molecular weight of 900,000 daltons and is a pentamer that the fetus can produce quite early in gestation.

464. The answer is C (2, 4). *(Davis, ed 2. pp 407-408, 415, 422.)* Kappa and lambda light chains are immunoglobulin precursors of Bence Jones (myeloma) proteins. Both light chains of each IgG molecule are of the same type, except when investigators have reoxidized reduced fragments to form hybrid antibodies. The Inv allotypic locus is manifest on kappa chains only; Inv(1, 2) is leucine-191, while Inv(3) is valine-191. Genes for kappa and lambda chains segregate independently.

465. The answer is B (1, 3). *(Davis, ed 2. pp 372-373.)* The ascending limb ("A") of the precipitin curve presented in the question represents the zone of antibody excess; in this zone, the supernatant solution would contain unreacted antibody. On the descending limb "C," or the antigen-excess zone, the supernatant solution contains excess free antigen. In a monospecific system, "B" designates the region of maximum precipitation and the supernatant solution is free to precipitate antibody and antigen. In a complex multispecific system, excess antigen or antibody molecules may be present at the point of maximum precipitate formation, because the optimal quantity of each antigen may be different.

466. The answer is C (2, 4). *(Davis, ed 2. pp 411, 416, 418.)* Immunoglobulin M and immunoglobulin A both have J chains in addition to their heavy and light chains. J chains appear to stabilize the immunoglobulins but are not necessary for maintaining their polymeric structure. The molecular weight of J chains is approximately 20,000 daltons.

467. The answer is A (1, 2, 3). *(Davis, ed 2. p 411.)* The Fab portions of immunoglobulins contain complete light chains and N-terminal fragments of heavy chains. The Fc portions contain C-terminal fragments of heavy chains but no light chains. Thus, antisera to Fab react to heavy and light chains, while antisera to Fc react only to heavy chains.

468. The answer is A (1, 2, 3). *(Davis, ed 2. pp 377-378, 410.)* Neither monovalent antigen nor monovalent antibody (Fab') can form a precipitin lattice. An antigen molecule containing closely repeating antigenic determinants (e.g., a polysaccharide or a multichained polymeric protein) can bind antibody to two determinants on a single particle; this "monogamous bivalency" inhibits precipitation. F(ab')$_2$ divalent antibodies can precipitate antigens, though they lack Fc portions.

469. The answer is A (1, 2, 3). *(Bellanti, p 163.)* A number of characteristics can be used to differentiate B and T lymphocytes. For example, Thy 1.1 and Thy 1.2, which are surface alloantigens, are found on T cells (as well as skin and fibroblast cells) but not on B cells. Cell-surface receptors that recognize the C3b component of complement are found primarily on B lymphocytes. B and T lymphocytes respond selectively to stimulation by phytohemagglutinin, concanavilin A, and pokeweed mitogen. Both B and T cells have antigenic memory.

470. The answer is B (1, 3). *(Bellanti, p 325.)* Serum sickness is a systemic immune disorder caused by small, soluble antigen-antibody aggregates circulating in the bloodstream. Deposition of these complexes in the renal glomerulus may lead to glomerular damage. Circulating nuclear autoantibodies and free nuclear antigens are responsible for the nephritis of systemic lupus erythematosus. Habitual phenactin ingestion can lead to papillary necrosis. The primary pathologic event in the hemolytic-uremic syndrome is occlusion of glomerular capillaries and renal arterioles by fibrin thrombi.

471. The answer is E (all). *(Davis, ed 2. pp 354-355.)* Haptens are substances that are not immunogenic in and of themselves. However, haptens can competitively inhibit the binding of a hapten-protein complex to an antibody that has been elaborated in response to that complex. Drug allergies involving haptenic determinants require sensitization to a modified form of the drug coupled to an endogenous protein (carrier).

472. The answer is A (1, 2, 3). *(Davis, ed 2. pp 460, 463.)* T cells bearing sheep-erythrocyte receptors may be removed from peripheral blood by formation of sheep-cell rosettes and subsequent sedimentation. Nylon fiber columns bind IgG-bearing B lymphocytes and allow T cells to pass. A cell separator also separates pools of immunoglobulin-bearing B cells and nonimmunoglobulin-bearing T cells. Anti-theta serum and complement may be used to prepare B cells by lysis of T cells; however, T cells are lost by this method.

473. The answer is E (all). *(Davis, ed 2. pp 419-420.)* Single amino-acid substitutions can give rise to antigenically distinct molecules. Hemoglobin, haptoglobin, glucose-6-phosphate dehydrogenase, and HL-A histocompatibility antigen all have allelic variants. Allotypes can be immunogenic in other individuals of the same species.

474. The answer is E (all). *(Davis, ed 2. pp 482-489.)* With repeated immunization, higher titers of all antibodies are observed, and as priming is repeated, the immune response recruits B cells of progressively greater affinity. As the affinity of antibody for a hapten-protein complex rises, cross-reactivity also rises, and the response becomes wider and wider in specificity. As the multiplicity of antigenic sites detected per reacting particle increases, the avidity increases. In addition to shifts in the class of immunoglobulin synthesized in response to an antigen (IgM to IgG), shifts also may occur in the idiotype of antibody.

475. The answer is E (all). *(Cave, Mt Sinai J Med NY 43 [1976]:795-829.)* A great many clinical conditions other than syphilis can give rise to a "reactive" or, more commonly, a "weakly reactive" serologic reading, which is actually a false positive. Leprosy, malaria, disseminated lupus erythematosus, and periarteritis nodosa are the most common of these conditions. False-positive reactions are relatively common with reagin or cardiolipin tests. If clinical signs of either syphilis or one of the common causes of these false positives are absent, then a treponemal serology technique should be used. In addition, the reagin tests should be repeated both immediately and several months later; other reagents or techniques should be used. If the treponemal work-up now proves negative, the other diagnoses should be pursued further.

476-480. The answers are 476-D, 477-B, 478-C, 479-A, 480-E. *(Bellanti, p 117. Jawetz, ed 13. pp 146-148.)* Immunoglobulins are divided into five classes: IgA, IgD, IgE, IgG, and IgM. Every immunoglobulin is made up of light and heavy polypeptide chains held together by disulfide bonds. Light chains are of two types, kappa and lambda, only one of which occurs in any given antibody molecule. On the other hand, each immunoglobulin class has its own distinct type of heavy chain.

The five classes of immunoglobulins can be differentiated by their molecular weights. IgA can exist as a monomer in serum (molecular weight 170,000, sedimentation coefficient 7S), a dimer, or a trimer in secretions (400,000, 11S). Except under the influence of reducing agents, IgM exists as a pentamer; its molecular weight is 900,000 and its sedimentation coefficient 19S. Existing only as monomers are IgD (180,000, 7S), IgE (190,000, 8S), and IgG (150,000, 7S).

481-485. The answers are: 481-D, 482-C, 483-D, 484-B, 485-D. *(Davis, ed 2. pp 506-509.)* Immunodeficiency disorders can be categorized according to whether the defect primarily involves humoral immunity (bone-marrow-derived, or B, lymphocytes) or cellular immunity (thymus-derived, or T, lymphocytes) or both. Swiss-type hypogammaglobulinemia, ataxia-telangiectasia, the Wiskott-Aldrich syndrome, and severe combined immunodeficiency disorders all involve defective B-cell and T-cell function. Infantile X-linked agammaglobulinemia is caused chiefly by deficient B-cell activity, whereas thymic hypoplasia is mainly a T-cell immunodeficiency disorder.

486-489. The answers are: 486-C, 487-A, 488-B, 489-D. *(Davis, ed 2. pp 378-379, 392-393, 396.)* The Farr technique of ammonium sulfate precipitation depends upon the differential precipitation of antibody and antigen-antibody complexes in a solution of high-salt concentration (40%). Unbound antigen must be soluble at this salt concentration. By the Farr technique, titration can be used to determine the antigen-binding capacity as well as the intrinsic association constants of antibody and antigen.

The Coombs technique is a routinely employed clinical procedure. Nonagglutinating, or incomplete, antibodies adsorbed onto red blood cells are detected by means of an antiglobulin. The Coombs test amplifies reactions that would otherwise be impossible to detect by immunofluorescence or hemagglutination.

The observation that much more antigen (toxin) is able to be neutralized by a given amount of antibody (antitoxin) when the antigen is added all at once instead of in divided portions was described first by Danysz, who studied the diphtheria toxin-antitoxin system. When a given amount of antigen is added to antibody in divided portions over a period of time, antigen-antibody complexes form that are irreversibly bound, leaving little, if any, free antibody to neutralize subsequently added antigen. The same phenomenon occurs in precipitation, where equivalence points may depend upon the manner in which antigen and antibody are combined.

Monogamous bivalency is the term applied by Karush to the nonprecipitating combination of high-affinity antibodies and antigenic determinants on the same particle. This phenomenon depends upon repetitive, closely spaced antigenic determinants on the same particle, with which a single antibody molecule may react. Thus, all of an antibody's binding sites are occupied by antigenic determinants from one molecule, making the cross-linking of antigen particles – and, therefore, precipitation – unlikely.

490-492. The answers are: 490-D, 491-C, 492-B. *(Davis, ed 2. pp 406, 419-425.)* Isotypes are determined by antigens of the major immunoglobulin classes found in all individuals of one species. In addition to heavy-chain isotypes of IgA, IgD, IgE, IgG, and IgM, two light-chain isotypes exist for kappa and lambda chains. Allotypes are differentiated by antigenic determinants that vary among individuals within a species and are recognized by cross-immunization of individuals in a species. Allotypes include the Gm marker of IgG and the Inv *or Km* marker of light chains. Idiotypes are antigenic determinants that appear only on the Fab fragments of antibodies and appear to be localized at the ligand-binding site; thus, anti-idiotype antisera may block reactions with the appropriate hapten. The carbohydrate side-chains of immunoglobulins are relatively nonimmunogenic. New determinants may be exposed after papain cleavage of immunoglobulins, but these determinants are not included in the classification of the native molecule.

allotype Am of IgA
Inv (Km) only on K chains
Kernd ozon λ are isotypes

493-497. The answers are: 493-B, 494-C, 495-D, 496-E, 497-A. *(Davis, ed 2. pp 425-427. Williams, ed 2. p 1093.)* Electrophoresis of human serum proteins identifies five distinct types: albumin, α_1-proteins, α_2-proteins, β-proteins, and gamma globulins. A normal electrophoretic profile appears below:

Albumin

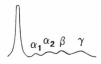

Many human diseases can be diagnosed, at least in part, on the basis of abnormal electrophoretic profiles. For example, absence of the second peak (α_1) is compatible with a diagnosis of α_1-antitrypsin deficiency in symptomatic individuals. A sharp and high gamma peak indicates the presence of a monoclonal gammopathy, such as multiple myeloma; on the other hand, a gamma peak that is diffusely elevated points to a polyclonal hypergammaglobulinemia. Complete absence of the gamma peak is associated with Swiss-type agammaglobulinemia.

498-500. The answers are: 498-C, 499-B, 500-B. *(Jawetz, ed 13. pp 159-160.)* The complement-fixation (CF) test is a two-stage test. The first stage involves the union of antigen with its specific antibody, followed by the fixation of complement to the antigen-antibody structure. In order to determine whether complement has been "fixed," an indicator system must be employed to determine the presence of free complement. Free complement binds to the complexes formed when red blood cells (RBC) are mixed with anti-RBC antibody; this binding causes lysis of the cells. Complement that has been "fixed" before addition of red blood cells and anti-RBC antibody cannot cause lysis.

Bibliography

Bibliography

Ames, G.F. "Resolution of bacterial proteins by polyacrylamide gel electrophoresis on slabs. Membrane, soluble, and periplasmic fractions." *Journal of Biological Chemistry* 249 (1974):634-644.

Bellanti, J.A. *Immunology II.* Philadelphia: Columbia Broadcasting System, W.B. Saunders Co., 1978.

Braude, A.I. *Antimicrobial Drug Therapy.* Major Problems in Internal Medicine Series, vol 8. Philadelphia: Columbia Broadcasting System, W.B. Saunders Co., 1976.

Briody, B.A., ed. *Microbiology and Infectious Disease.* New York: McGraw-Hill Book Co., Blakiston Publications, 1974.

Brown, H.W. *Basic Clinical Parasitology.* 4th ed. New York: Prentice-Hall Inc., Appleton-Century-Crofts, 1975.

Burrows, W. *Textbook of Microbiology.* 20th ed. Philadelphia: Columbia Broadcasting System, W.B. Saunders Co., 1973.

Cahill, K.M. *Tropical Diseases: A Handbook for Practitioners.* Westport: Technomic Publishing Co., 1976.

Cave, V., and Nikitas, J. "Venereal Disease – clinical and laboratory diagnoses." *Mount Sinai Journal of Medicine, New York* 43 (1976):795-829.

Copenhaver, W.M.; Bunge, R.P.; and Bunge, M.P. *Bailey's Textbook of Histology.* 17th ed. Baltimore: Williams & Wilkins Co., 1978.

Davis, B.D., et al. *Microbiology.* 2nd ed. New York: Harper & Row Publishers Inc., 1973.

Faust, E.C.; Russell, P.; and Jung, R. *Craig and Faust's Clinical Parasitology.* 8th ed. Philadelphia: Lea & Febiger, 1970.

Finegold, S.M.; Martin, W.J.; and Scott, E. *Bailey and Scott's Diagnostic Microbiology.* 5th ed. St. Louis: C.V. Mosby Co., 1978.

Fraser, D.W.; Tsai, R.R.; Orenstein, W.; et al. "Legionnaires' disease: Description of an epidemic of pneumonia." *New England Journal of Medicine* 297 (1977):1189-1197.

Fujimoto, W.Y.; Subak-Sharpe, J.H.; and Seegmiller, J.E. "Hypoxanthine-guanine phosphoribosyltransferase deficiency: Chemical agents selective for mutant or normal cultured fibroblasts in mixed and heterozygote cultures." *Proceedings of the National Academy of Sciences of the United States of America* 68 (1971):1516-1519.

Gell, P.G.; Coombs, R.R.; and Lachman, P.J., eds. *Clinical Aspects of Immunology.* 3rd ed. Philadelphia: J.B. Lippincott Co., 1975.

Goodman, L.S., and Gilman, A. *The Pharmacological Basis of Therapeutics.* 5th ed. New York: Macmillan Publishing Co. Inc., 1975.

Harvey, A.M., et al. *The Principles and Practice of Medicine.* 19th ed. New York: Prentice-Hall Inc., Appleton-Century-Crofts, 1976.

Hoeprich, P.D., ed. *Infectious Diseases.* 2nd ed. New York: Harper & Row Publishers Inc., 1977.

Ishikawa, H.; Bischoff, R.; and Holtzer, H. "Formation of arrowhead complexes with heavy meromyosin in a variety of cell types." *Journal of Cell Biology* 43 (1969):312-328.

Jawetz, E.; Melnick, J.L.; and Adelberg, E.A. *Review of Medical Microbiology.* 13th ed. Los Altos: Lange Medical Publications, 1978.

Joklik, W.K., and Willett, H.P. *Zinnser Microbiology.* 16th ed. New York: Prentice-Hall Inc., Appleton-Century-Crofts, 1976.

Lehninger, A.L. *Biochemistry: The Molecular Basis of Cell Structure and Function.* 2nd ed. New York: Worth Publishers Inc., 1975.

Lennette, E.H.; Spaulding, E.H.; and Truant, J.P., eds. *Manual of Clinical Microbiology.* 2nd ed. Washington: American Society for Microbiology, 1974.

McDade, J.E.; Shepard, C.C.; Fraser, D.W.; et al. "Legionnaires' disease: Isolation of a bacterium and demonstration of its role in other respiratory diseases." *New England Journal of Medicine* 297 (1977):1197-1203.

Nicolson, G.L., and Singer, S.J. "The distribution and asymmetry of mammalian cell surface saccharides utilizing ferritin-conjugated plant agglutinins as specific saccharide stains." *Journal of Cell Biology* 60 (1974):236-248.

Thorn, G.W., et al., eds. *Harrison's Principles of Internal Medicine.* 8th ed. New York: McGraw-Hill Book Co., Blakiston Publications, 1977.

Tobey, R.A. "Arrest of Chinese hamster cells in G2 following treatment with the anti-tumor drug bleomycin." *Journal of Cellular Physiology* 79 (1972):259-266.

Tobey, R.A., and Ley, K.D. "Isoleucine-mediated regulation of genome replication in various mammalian cell lines." *Cancer Research* 31 (1971):46-51.

Van Gool, A.P., and Nanninga, N. "Fracture faces in the cell envelope of *Escherichia coli.*" *Journal of Bacteriology* 108 (1971):474-481.

Volk, W.A. *Essentials of Medical Microbiology.* Philadelphia: J.B. Lippincott Co., 1978.

Watson, J.D. *Molecular Biology of the Gene.* 3rd ed. Reading: Addison-Wesley Publishing Co., Benjamin-Cummings Publishing Co., 1976.

Williams, W.J., et al. *Hematology.* 2nd ed. New York: McGraw-Hill Book Co., Blakiston Publications, 1977.

Youmans, G.P., and Paterson, P.Y. *The Biological and Clinical Basis of Infectious Diseases.* Philadelphia: Columbia Broadcasting System, W.B. Saunders Co., 1975.

NOTES

NOTES

NOTES